SUSTAINABLE SOIL MANAGEMENT

SUSTAINABLE SOIL MANAGEMENT

Edited by

Deirdre Rooney, PhD

Apple Academic Press

TORONTO NEW JERSEY

Apple Academic Press Inc. 3333 Mistwell Crescent Oakville, ON L6L 0A2 Canada	Apple Academic Press Inc. 9 Spinnaker Way Waretown, NJ 08758 USA

Printed in the United States of America on acid-free paper

International Standard Book Number-13: 978-1-926895-21-5 (Hardcover)

Library of Congress Control Number: 2012951954

Library and Archives Canada Cataloguing in Publication

Sustainable soil management/edited by Deidre Rooney.

Includes bibliographical references and index.
ISBN 978-1-926895-21-5
1. Soil management. 2. Soil management--Environmental aspects. I. Rooney, Deirdre

S591.S98 2013 631.4 C2012-906397-5

Apple Academic Press also publishes its books in a variety of electronic formats. Some content that appears in print may not be available in electronic format. For information about Apple Academic Press products, visit our website at **www.appleacademicpress.com** and the CRC Press website at **www.crcpress.com**

About the Editor

Deirdre Rooney, PhD

Deirdre Rooney is currently employed in the agriculture department at Askham Bryan College, York (UK), as a lecturer in environmental and biological sciences. She graduated from the University College Dublin, Ireland, with a BSc (Honours) in botany and subsequently completed a PhD in the field of microbial ecology, investigating soil nitrogen dynamics in grassland ecosystems. She has since carried out further research in soil microbiology, with a specific focus on plant-soil interactions, including those of soil bacteria and mycorrhizal fungi.

Contents

List of Contributors

J. Richard Alldredge
Department of Statistics, Washington State University, Pullman Washington, United States of America

Preston K. Andrews
Department of Horticulture and Landscape Architecture, Washington State University, Pullman, Washington, United States of America

Martin A. N. Anikwe
Department of Agronomy and Ecological Management, Faculty of Agriculture and Natural Resources Management, Enugu State University of Science and Technology, P.M.B. 01660 Enugu, Nigeria

Sébastien Barot
Bioemco (UMR 7618), Site Ecole Normale Supérieure, 46 rue d'Ulm, 75230 Paris cedex 05, France

Rainer W. Bussmann
William L. Brown Center, Missouri Botanical Garden, St. Louis, Missouri, United States of America

Lynne Carpenter-Boggs
Center for Sustaining Agriculture and Natural Resources, Washington State University, Pullman, Washington, United States of America

Nicholas J. W. Clipson
Microbial Ecology Group, School of Biology and Environmental Science, University College Dublin, Belfield, Dublin 4, Ireland

Neal M. Davies
Department of Pharmaceutical Sciences, Washington State University, Pullman, Washington, United States of America

Thibaud Decaëns
Laboratoire d'Ecologie, UPRES EA 1293 ECODIV, FED SCALE, UFR Sciences et Techniques, Université de Rouen, 76821 Mont Saint Aignan cedex, France

Tyler A. Demers
School of Environmental Sciences, University of Guelph, 50 Stone Road East, Guelph, Ontario, Canada

Andrew K. Evers
School of Environmental Sciences, University of Guelph, 50 Stone Road East, Guelph, Ontario, Canada

Luise Giani
Department of Soil Science C-v-O, University of Oldenburg, Oldenburg, Germany

Deirdre B. Gleeson
Soil Biology Group, School of Earth and Environment (M087), The University of Western Australia, 35 Stirling Highway, WA 6009 Crawley, Australia

Andrew M. Gordon
School of Environmental Sciences, University of Guelph, 50 Stone Road East, Guelph, Ontario, Canada

Sergey V. Goryachkin
Department of Soil Science C-v-O, University of Oldenburg, Oldenburg, Germany

Pascal Jouquet
Bioemco (UMR 7618)—IWMI, SFRI, Dong Ngac, Tu Liem, Hanoï, Vietnam

Radha D. Kale
Centre for Scientific Research and Advanced Learning, Mount Carmel College, Bangalore, Karnataka 560 052, India

Olga Kalinina
Department of Soil Science C-v-O, University of Oldenburg, Oldenburg, Germany

Natchimuthu Karmegam
Department of Biotechnology, VMKV Engineering College, Vinayaka Missions University, Periya Seeragapadi, Salem, Tamil Nadu 636 308, India

Nina A. Karavaeva
Institute of Geography, Russian Academy of Sciences, Moscow, Russia

Nabla M. Kennedy
Microbial Ecology Group, School of Biology and Environmental Science, University College Dublin, Belfield, Dublin 4, Ireland

Munesh Kumar
Department of Forestry, HNB Garhwal University, Srinagar Garhwal, Uttarakhand, India

Zhengpeng Li
SAIC, contractor to US Geological Survey (USGS) Center for Earth Resources Observation and Science, Sioux Falls, South Dakota, United States of America

Shuguang Liu
SAIC, contractor to US Geological Survey (USGS) Center for Earth Resources Observation and Science, Sioux Falls, South Dakota, United States of America

Kam-Rigne Laossi
Laboratoire d'Ecologie, UPRES EA 1293 ECODIV, FED SCALE, UFR Sciences et Techniques, Université de Rouen, 76821 Mont Saint Aignan cedex, France

Thomas R. Loveland
US Geological Survey (USGS) Center for Earth Resources Observation and Science, Sioux Falls, South Dakota, United States of America

Dmitriy I. Lyuri
Institute of Geography, Russian Academy of Sciences, Moscow, Russia

Swati Pattnaik
Department of Ecology and Environmental Sciences, Pondicherry University, Puducherry 605 014, India

Jennifer S. Powers
Department of Ecology, Evolution and Behavior, University of Minnesota, St. Paul, Minnesota, United States of America; Department of Plant Biology, University of Minnesota, St. Paul, Minnesota, United States of America; Department of Soil, Water and Climate, University of Minnesota, St. Paul, Minnesota, United States of America

M. Vikram Reddy
Department of Ecology and Environmental Sciences, Pondicherry University, Puducherry 605 014, India

Jennifer R. Reeve
Department of Plants, Soils and Climate, Utah State University, Logan, Utah, United States of America

John P. Reganold
Department of Crop and Soil Sciences, Washington State University, Pullman, Washington, United States of America

Deirdre C. Rooney
Microbial Ecology Group, School of Biology and Environmental Science, University College Dublin, Belfield, Dublin 4, Ireland

Carolyn F. Ross
School of Food Science, Washington State University, Pullman, Washington, United States of America

Emma J. Sayer
Department of Plant Sciences, University of Cambridge, Cambridge, United Kingdom; Smithsonian Tropical Research Institute, Balboa, Ancon, Panama, Republic of Panama

Christopher W. Schadt
Biosciences Division, Oak Ridge National Laboratory, Oak Ridge, Tennessee, United States of America

Mehraj A. Sheikh
Department of Forestry, HNB Garhwal University, Srinagar Garhwal, Uttarakhand, India

Zhengxi Tan
SAIC, contractor to US Geological Survey (USGS) Center for Earth Resources Observation and Science, Sioux Falls, South Dakota, United States of America

Edmund V. J. Tanner
Department of Plant Sciences, University of Cambridge, Cambridge, United Kingdom

David J. Tenenbaum
National Association of Science Writers, Berkeley, California, United States of America

Naresh V. Thevathasan
School of Environmental Sciences, University of Guelph, 50 Stone Road East, Guelph, Ontario, Canada

Jizhong Zhou
Department of Botany and Microbiology, Institute for Environmental Genomics, University of Oklahoma, Norman, Oklahoma, United States of America

List of Abbreviations

ABTS	2,2-Azino-bis-(3-ethylbenzthiazoline-6-sulfonicacid)
AMO	Ammonia monooxygenase
AMPS	Automated stochastic parameterization system
ANOSIM	Analysis of similarity
ANOVA	Analysis of variance
AOB	Ammonia-oxidizing bacteria
AT	Atmospheric temperature
ATM	Actual tillage management
BSA	Bovine serum albumin
CDM	Clean development mechanism
CER	Carbon emission reduction
CER	Certified emission reduction
CFUs	Colony forming units
CON	Conventional
CRP	Conservation reserve program
CT	Conventional tillage
CTIC	Conservation technology information center
DRI	Dietary reference intakes
DTT	DL-dithiothreitol
EC	Electrical conductivity
FC	Folin-Ciocalteu
FIA	Forest inventory and analysis data
GARS	Guelph agroforestry research station
GEMS	General ensemble biogeochemical modeling system
GIS	Geographic information system
GRSP	Glomalin-related soil protein
HAA	Hydrophilic antioxidant activities
IBI	International Biochar Initiative
INFOODS	International network of food data systems
IOP	Input/output processor
IRGA	Infra-red gas analyzer
IS	Internal standard
JFD	Joint frequency distribution
LAA	Lipophilic antioxidant activities
LSD	Least significant difference
MSS	Multispectral scanner
MW	Market waste
NEE	Net ecosystem exchange
NPP	Net primary productivity
NT	No-till
OC	Organic carbon

OM	Organic matter
ORG	Organic
OUT	Operational taxonomic unit
PCO	Principal coordinate analysis
PERMANOVA	Permutational multivariate analysis of variance
POM	Particulate organic matter
RT	Reduced tillage
SIs	Signal intensities
SM	Soil moisture
SNR	Signal-to-noise ratio
SOC	Soil organic carbon
SOM	Soil organic matter
SPSS	Statistical package for the social sciences
ST	Soil temperature
STATSGO	State soil geographic data base
TA	Titratable acidity
TAA	Total antioxidant activity
TBI	Tree-based intercropping
TC	Tungsten compounds
TEL	Tetra ethyl leads
TIN	Total inorganic N
TM	Thematic mapper
TOC	Total organic carbon
TRF	Terminal restriction fragment
TRFLP	Terminal restriction fragment length polymorphism
UNCCD	United Nations convention to combat desertification
USDA	United States department of agriculture
USGS	U.S. geological survey
VC	Vermicompost
WRB	World reference base
WSU	Washington State University

Introduction

Soil is one of our most precious natural resources, providing a range of ecosystem functions and services, while supporting huge biodiversity. Such functions and services include plant growth, nutrient cycling, decomposition of organic matter, breakdown of pollutants, and carbon storage. Despite its importance, many questions still remain unanswered about soil function and its sustainable management in the 21st century.

In these chapters, the focus is on soil management, with an emphasis on the potential for sustainable soil management for agriculture and the environment. Changing land-use practices and the role of soil biological diversity has been a major focus of soil science research over the past couple of decades, a trend that is likely to continue as environmental issues such as climate change and rising greenhouse gas levels, and concerns regarding food production for a growing population, become increasingly debated topics. Content presented in these chapters suggests a holistic approach to soil management is required.

Part I of this book, "Land Use Effects on Soil Carbon Storage," considers how a range of factors influence carbon sequestration in soils. With a significant amount of global carbon stocks held in soil, land management practices have the potential to mitigate rising atmospheric CO_2 levels through carbon sequestration. Agricultural practices have a significant impact on the quantity of carbon held in soils, as indicated by Anikwe in chapter 1. This chapter presents evidence suggesting that conventional tillage systems can lose up to 70% more soil carbon than long-term forest or grassland ecosystems. Deforestation of land for crop production may therefore have detrimental consequences for global carbon storage.

In chapter 2, Tan et al. simulated the impacts of tillage regimes on soil carbon stocks in the Northwest Great Plains (US) using biogeochemical modeling and concluded that no-tillage systems released less carbon than conventional tillage. However, the establishment of baseline soil organic carbon levels was critical in estimating the potential of different agricultural soils to mitigate carbon emissions. In addition, these authors also revealed the importance of considering local and regional scale cropping regimes. Both of these chapters indicate that much variability exists regarding carbon mitigation in agricultural soils and a deeper understanding of climate influences, soil type, existing carbon stock, and land management strategies is needed.

A new concept for carbon storage is described by Tenenbaum in chapter 3: the potential of biochar, a non-fuel substance similar to charcoal, for long-term soil carbon storage. Incorporation of biochar into soils may promote soil fertility and productivity while enhancing soil microbial function. Although further long-term trials are required, evidence suggests that the stable and recalcitrant properties of biochar may offer a long-term solution to soil carbon sequestration.

Increasing atmospheric CO_2 levels is likely to increase aboveground plant productivity as suggested by Sayer et al. in chapter 4, and may result in greater quantities of litterfall, a major pathway for carbon locked in vegetation to reach the soil. Sayer et al. present data proposing that increased litterfall could accelerate losses of soil organic carbon via heterotrophic respiration, due to a priming effect. Another study investigating forest ecosystems in relation to soil carbon storage is that of Sheikh et al. in chapter 5. Sheikh et al. found a negative correlation between altitude and the ability of soils to store organic carbon, in both temperate deciduous and subtropical coniferous forest ecosystems. Both these chapters serve to emphasize the range of factors that may affect soil carbon sequestration.

In chapter 6, Kalinina et al. investigated the recovery of soil carbon stocks in soils previously used for agricultural production. Exploring soil chronosequences spanning 170 years, the authors indicate that soil organic carbon in the mineral layers remained relatively stable over this period, but carbon sequestration increased in the developing organic surface layers. In conclusion, agricultural practices can have long-lasting effects on soil carbon storage potential.

Part II of this book, "Soil Biotic Interactions," presents chapters investigating the interactions between soil properties, plant species, and the soil biota. Soil biota, both macro and micro, play important roles in soil function particularly in processes such as decomposition of organic material, improvement of soil structure, and nutrient cycling. Exploiting the ability of the soil biota to perform these functions may contribute to holistic and sustainable soil management in the future.

In chapter 7, Kale and Karmegam describe the significance of earthworms communities in soil nutrient cycling and emphasize the potential for earthworms in the production of vermicompost, a valuable soil conditioner and fertilizer. The nutritional qualities of different vermicomposts were investigated by Pattnaik and Reddy in chapter 8. This chapter reports that both the type of input waste (either vegetable waste or floral waste) plus the species of earthworm present determined the end quality of the vermicompost. These findings may have important implications for waste management and recycling of organic waste.

Furthermore, earthworms serve as bioindicators if soil quality and their activity may be closely linked to microbial activity. In chapter 9, Evers et al. describe one such linkage between earthworms and denitrifier bacteria, the latter being responsible for nitrous oxide (N_2O) production in soils. The earthworm gut is anoxic, and represents an ideal environment for denitrification to occur. Chapter 9 also presents evidence of increased nitrous oxide production at higher earthworm densities, although this result was also dependent on soil water content. Understanding such soil interactions will become increasingly more important for future soil management.

The interactions between agricultural practices, soil properties, and soil microbial communities are further discussed in chapter 10 by Rooney et al., who present data on the effects of agricultural practices on key soil microbial communities in unimproved and improved grassland soils. Oxidation of ammonia is a key step in nitrification (the conversion of ammonia to nitrate via nitrite) and one of the major groups of microbes that are involved in this process are the ammonia-oxidizing bacteria. Using molecular profiling of ammonia oxidizing bacterial communities, the authors concluded that interactions between nitrogen additions and plant species are key drivers in ammonia-oxidising bacterial community structure.

Further interactions between earthworms and soil properties on plant growth are reported in by Laossi et al in chapter 11. In this chapter, the authors describe a range of soil characteristics that influence plant-earthworm interactions, such as organic matter content, soil texture, and soil nutrient status. Interestingly, the authors concluded that different soil types promote different plant-earthworm relationships, with sandy soils generally yielding the most positive results in terms of positive earthworm effects on plant growth. Exploitation of such biotic interaction may have important consequences for alternative agricultural methods such as organic or low-input farming, which generally support higher levels of soil biota.

In the final chapter, in an investigation into fruit and soil quality under organic and conventional systems, Reganold et al. found that organically produced strawberries had longer shelf life, greater dry matter and greater nutritional value than those grown in conventional systems. Furthermore, the research indicated that organically managed soils had higher genetic diversity for processes such as nitrogen fixation and pesticide degradation, leading to the conclusion that organic soils are more resilient and of better quality than conventionally managed soils.

Together, all these chapters highlight some of the contemporary issues concerning land management, nutrient cycling, and soil carbon storage. Further research into these important areas will enhance our understanding of key functional processes in soils, potentially leading to a more holistic approach to sustainable land management in the future.

— **Deirdre Rooney, PhD**

PART I

LAND USE EFFECTS ON SOIL CARBON STORAGE

1 Carbon Storage in Southeastern Nigerian Soil

Martin A. N. Anikwe

CONTENTS

1.1 INTRODUCTION

Changes in agricultural practices—notably changes in crop varieties, application of fertilizer, manure, rotation, and tillage practices—influence how much and at what rate carbon is stored in or released from soils. Quantification of the impacts of land use on carbon stocks in Sub-Saharan Africa is challenging because of the spatial heterogeneity of soil, climate, management conditions, and due to the lack of data on soil carbon pools of most common agroecosystems. This chapter provides data on soil carbon stocks that were collected at 10 sites in southeastern Nigeria to characterize the impact of soil management practices.

The highest carbon stocks, 7906–9510 gC m^{-2} were found at the sites representing natural forest, artificial forest, and artificial grassland ecosystems. Continuously, cropped and conventionally tilled soils had about 70% lower carbon stock (1978–2822 gC m^{-2}). Thus, the soil carbon stock in a 45 year old *Gmelina* forest was 8987 gC m^{-2}, whereas the parts of this forest that were cleared and continuously cultivated for 15 years, had 75% lower carbon stock (1978 gC m^{-2}). The carbon stock of continuously cropped and conventionally tilled soils was also 25% lower than the carbon stock of the soil cultivated by use of conservation tillage.

Introducing conservation tillage practices may reduce the loss of soil carbon stocks associated with land conversion. However, the positive effect of conservation tillage is not comparable to the negative effect of land conversion and may not result in significant accumulation of carbon in southeastern Nigeria soils.

Soil organic carbon (SOC) is a large and active pool, containing roughly twice as much carbon as the atmosphere and 2.5 times as much as the biota. Carbon sequestration is the facilitated redistribution of carbon from the air to other pools. This would reduce the rate of atmospheric CO_2 increase, thereby mitigating global warming [1, 2].

The amount of carbon sequestered at a site reflects the long-term balance between influx and efflux of carbon. Recent concerns with rising atmospheric levels of CO_2 have stimulated interest in C flow in terrestrial ecosystems and the latter's potential for increased soil carbon sequestration [3]. Carbon enters the soil as roots, litter, harvest residues, and animal manure. It is stored primarily as soil organic matter (SOM). The density (w/v) of carbon is highest near the surface, but SOM decomposes rapidly, releasing CO_2 to the atmosphere. Some carbon becomes stabilized, especially in the lower part of the profile. However, in many areas, agricultural and other land use activities have upset the natural balance in the soil carbon cycle, contributing to an alarming increase in carbon release [4, 5]. Since the current rise in atmospheric CO_2 is thought to be mitigated in part by carbon sequestration in agricultural soils [4], interest has increased in the possible impacts of various agricultural management practices on SOM dynamics [6].

Agricultural and other land use practices have a significant influence on how much carbon can be sequestered and how long it can be stored in the soil before it is returned to the atmosphere. The best strategies focus on the protection of SOC against further depletion and erosion, or the replenishment of depleted carbon stocks through certain management techniques [2]. In either case, the keys to successful soil carbon sequestration are increased plant growth and productivity, increased net primary production, and decreased decomposition [2]. Similarly, conversion of marginal arable land to forestry or grassland can rapidly increase soil carbon sequestration. For example, analysis of long-term crop experiments indicated that increasing crop rotation complexity increased SOC sequestration by 20 gC m^{-2} yr^{-1}, on average [7]. In long-term experiments in Canada, the SOC sequestration rates were 50–75 gC m^{-2} yr^{-1} in well-fertilized soils with optimal cropping [8]. By contrast, long-term experiments in the Northern Great Plains (US) have shown that fertilizer N increased crop residue returns to the soil, but generally did not increase SOC sequestration [9]. Ogunwale and Raji [8] found that after 45 years of cow dung and NPK treatments to a soil in Samaru Northern Nigeria, SOC content in the unamended soil was 1.81 tC ha^{-1} or 10 gC m^{-2} between 1977 and 1995. In the same period of 45 years, the use of continuous NPK application resulted in only slight increase in SOC (3%) over the unamended soil while manure with NPK gave 115% more SOC. They found that the rate of SOC sequestration during fallow period in their experiment was approximately 400% more than the rates under continuous cultivation.

Timing and intensity of tillage also must be taken into account in the design of best management practices for maximizing SOC sequestration [10-12].

In most of Africa including Nigeria, research on quantification of carbon stored in the soil is proceeding slowly. Thus, data on soil C pools are lacking for most common agroecosystems. It is important to note that data collected from tropical environments are used in estimating total world carbon sequestration potential because differences in edaphoclimatic conditions and soil management practices influence the storage of carbon in the soil. For example, with the exception of histosols that have 13–27% SOM (by weight) [13], average SOM contents of soils in Sub-Saharan Africa range (between) 0.5–3.0%, whereas temperate Europe and America soils record up to 10–13% soil organic matter. Quantifying changes in soil C is a difficult task. Annual changes per year are small compared to C already present and its spatial variability can be very large [14]. Thus, reliable estimates of C change depend on sampling randomly at test sites over many years or by sampling at specific locations, repeatedly over time [15].

African countries are unlikely to engage in soil carbon sequestration unless there are clear local economic and societal benefits. Therefore, it is essential to estimate all potential costs and benefits related to the various management options. Large-scale adoptions of ecologically sound land use practices are likely to be the most cost effective and environmentally friendly option to increase soil carbon sequestration in Africa [2]. In addition, a correct measurement and verification of carbon sequestration potential of soils in Sub-Saharan Africa would enable the zone to participate in the clean development mechanism (CDM), proposed in study of the Kyoto Protocol to the United Nations Framework convention on climate change. This will allow developing countries to sell or trade project-based carbon credits, such as carbon emission reduction (CER) credits, to or with industrial countries, if adopted. The CER credits could provide an incentive for participation in climate change mitigation and cover the costs that African participants will encounter when engaging in carbon sequestration [2].

The objective of this work is to assess quantitatively, the effect of different soil management practices on SOC sequestration.

1.2 METHODS

1.2.1 Site Description

The soil samples were collected from 10 sites in different parts of Southeastern Nigeria. The differences in management practices and edaphoclimatic properties guided choice of different sites. Southeastern Nigeria stretches from 04°15'N to 07°00'N and between 05°34'E and 09°24'E, has a total area of approximately 78,612 km^2 [23]. Mean annual temperature ranges 27–32°C. The soils of the zone have isohyperthermic temperature regime and receive average annual rainfalls of 1600–4338 mm [23].

1.2.2 Observations and Data Collection

The soil samples used for the experiment were collected from 10 sites representing:

(a) Forests:

- (i) An artificial forest established by Forestry Department in 1962.
- (ii) A natural undisturbed forest (sacred land) that is more than 80 years old.
- (iii) An artificial *Gmelina arborea* forest established 30 years ago by the Forestry Department.

(b) Grassland:
  (i) Artificial grassland (golf course) established in 1934.

(c) Arable land:
  (i) Plot conventionally tilled with traditional hoes, planted with cassava/vegetables/maize intercrop with 2 year fallow period in 10 years.
  (ii) Plot conventionally tilled with traditional hoes, continuously cropped with maize/cassava/yam intercrop for 10 years.
  (iii) Plot conventionally tilled, unmulched, cropped to maize and cassava for 12 consecutive years.
  (iv) Land adjacent to the artificial forest cropped continuously for 15 years with cassava, yam, pulses, and vegetables in a mixed culture.
  (v) University Research plot, fallowed for 2 years and managed under conservation tillage for 3 years.
  (vi) Farmers plot, conventionally tilled, planted with cassava/maize/vegetables and used for 12 years.

The details of site number, location, soil classification, and land use history are presented in Table 1.

An initial (reconnaissance) survey was carried out in the 10 sites selected for the study to establish sampling points. Nine representative sampling points were chosen in each selected site using the free survey approach (observation points that are representative of the site are chosen by the surveyors based on personal judgment and experience) [24]. Three sampling depths (0–5, 5.1–15, and 15.1–30 cm) were used for the study. At each depth, nine undisturbed core samples and nine auger samples were collected for laboratory analysis.

The samples were collected at the end of the harvesting season in October when bulk density of tilled cropped fields had reverted to their pre-tillage conditions (because soil bulk density measurements are used for calculating carbon stocks) [17]. In cultivated plots, samples were collected randomly inside the rows. Auger samples were collected using a hand-pushed auger (Push Probe, 23 mm diameter). The core samples were collected using open-faced coring tube (area, 19.5 cm^3 and height, 5 cm from Eijkelkamp Agrisearch Equipment) at the three selected depths. Roots, twigs, and leaves were manually removed from auger samples and the samples air-dried at ambient temperature for 72 hr and subsequently sieved (using 2 mm sieves). The core samples were analyzed and mean results from each depth used whereas auger samples collected at a specific depth, were mixed and composite sub-samples (from each depth) used for further analyses.

The carbon stock in each agro ecological system was calculated with the formula = C (%)/100 × soil bulk density × area (1 ha) × soil depth.

1.2.3 Laboratory Methods

The samples were analyzed in the research laboratory of the department of soil science, University of Nigeria, Nsukka, for bulk density, gravimetric water content, organic carbon content, total nitrogen, soil pH, and particle size distribution. Bulk density was analyzed by core method [25]. Organic carbon was determined by the Walkley-Black procedure [26] and total nitrogen was by the Macro-Kjeldahl method

TABLE 1 Location, classification, and land use history of the 10 sites used for the study.

Site Number	Location/Annual rainfall	Soil Classification	Land use history
1	Abakaliki I 6°19'N,8°06'E 2069 mm	Aquept Flood Plain	Conventionally tilled with traditional hoes, planted with Cassava {*Manihot esculenta*}/vegetables (Amaranth {*Amaranthus hybridus*}, Okra {*Albemoschus esculentus*},Waterleaf {*Talinum triangulare*})/maize {*Zea mays*} inter-crop with 2-year fallow period in 10 years, no fertilization, crop residues not removed.
2	Enugu I 6°27'N,7°29'E 1792 mm	Typic Paleustult Mid-slope	Conventionally-tilled with traditional hoes, continuously-cropped with maize/cassava/yam {*Dioscorea rotundata*} intercrop for ten years. NPK 15:15:15 fertilizers used at low doses (30–50 kg ha^{-1}), crop residues left in the field.
3	PortHarcourt 4°46'N,7°01'E 2450 mm	Typic Paleustult Floodplain	Conventionally-tilled, unmulched, cropped to maize and cassava for 12 consecutive years, fertilized with low dose (30–50 kg ha^{-1}) of NPK 15:15:15 fertilizer, crop residues left in the field.
4	Enugu II 06° 27'N,7°32'E 1792 mm	Typic paleustult Midslope	Artificial forest established by Forestry Department in 1962. Planted with *Gmelina arborea* and *Tectona grandis* (Teak).
5	Enugu III 06° 27'N,7°32'E 1792 mm	Typic paleustult Midslope	Adjacent land near the artificial forest cropped-continuously for 15 years with cassava, yam, pulses and vegetables in a mixed culture. No fertilization and crop residues not removed.
6	Ihe,Awgu I 06° 30'N,7°15'E 1752 mm	Typic paleudult Toeslope	Natural undisturbed forest (sacred land). Had existed for more than 80 years. People are forbidden entry. Hunting of animals/games and cutting of trees/fetching of firewood not allowed.
7	Enugu IV 06° 27'N,7°25'E 1750 mm	Typic Paleudult Midslope	Artificial grassland (golf course) established in 1934. Mainly made up of *Paspalum notatum*, *Axonopus compressus* and *Cyperus rotundus*. Regularly cut and fertilized with N:P:K 15:15:15 fertilizer.

TABLE 1 *(Continued)*

Site Number	Location/Annual rainfall	Soil Classification	Land use history
8	Abakaliki II 6°04'N,8°65'E 2069 mm	Typic Haplaudult Crest	Artificial *Gmelina arborea* forest established 30 years ago. The alleys between the trees are currently cropped with different food crops(cocoyam, yam, cowpea, maize) by urban farmers. Municipal wastes used for fertilization, conventionally-tilled.
9	Nsukka, I 6°52'N,7°24'E 1700 mm	Typic Paleustult Footslope	University Research plot, fallowed for two years, conservation tillage, planted with maize and groundnut, fertilization with NPK 15:15:15 at 90 kg ha^{-1} and poultry droppings at 10 Mg ha^{-1} for three years.
10	Ihe,Awgu II 06° 30'N,7°05'E 1752 mm	Typic Paleudult Toeslope	Farmers plot, conventionally-tilled, planted with cassava/maize/vegetables (fluted pumpkins, cabbage, cowpea, Amaranth) with 2-year fallow interval, NPK 15:15:15 fertilizers used at low doses (30-50 kg ha^{-1}), crop residues left in the field, farm managed as stated for 12 years.

Anikwe, Carbon Balance and Management 2010; 5:5.

[27], whereas soil pH on a saturated sample was determined in soil electrolyte (0.01 M $CaCl_2$) suspension using a glass electrode pH meter (Digital pH meter, Accumet Model AR15, Fisher Scientific). Particle size distribution was determined using the pipette method of Gee and Orr [28].

1.3 DISCUSSION

High coefficients of variability in organic carbon and total N content were observed for SOC (SOC: 53–55%) and 178–184% for total N. High variability in SOC and total N content may indicate soil properties that are mostly impacted on the short to medium term by changes in soil management practices. Although measured values of bulk density even among the same soil vary considerably because densification of surface soil is caused by many factors *viz.* trafficking by humans and animals, wetting and drying cycles in soils, raindrop impact energy, and so on [17], the low coefficient of variation observed among the different soils used for the study especially in cultivated plots, may come from the fact that samples were collected at the end of the harvesting season when soil recompaction after tillage may have occurred. However, bulk density values are most useful in carbon sequestration studies for the calculation of total quantities of carbon sequestered at a particular time and soil depth. Krull et al. [18] stated that almost all organic carbon in soil is located within pores between mineral particles either as discrete particles or as molecules adsorbed onto the surfaces of these mineral particles. Soil architecture can influence biological stability of organic materials through its effects on water and oxygen availability, entrapment and isolation from decomposers, and through the dynamics of soil aggregation.

The highest SOC content was found in natural undisturbed forest, whereas lowest SOC was observed in conventionally tilled, continuously cropped plots. The studies [17] and [19] showed that tillage adversely affects carbon storage in the soil. However, although sites 3 and 9 were continuously tilled plots, their SOC contents were considerably high (2.3 and 1.06% in the 0–5 cm soil layer, respectively) when compared to sites either under grassland or forests probably because site 3 is a natural floodplain (see Table 1) whereby it seemed that enrichment of SOC occurred during yearly flooding. For site 9 in particular, the plot was managed under conservation tillage with annual addition of 20 t ha^{-1} of poultry droppings for 3 years. These may have drastically increased SOC of sites 3 and 9. Differences in SOC content of site 4 (Artificial *Gmelina arborea* forest) and site 5 (adjacent CT-CC plot) show that land clearing and continuous cultivation drastically reduce SOC. Bationo et al. [20] studying SOC dynamics, functions and management in West African agroecosystems reported rapid decline of SOC levels with continuous cultivation. For the sandy soils, they found that average annual losses may be as high as 4.7% whereas with sandy loam soils, losses were lower with an average of 2%. They postulated that total system carbon in different vegetation and land use types indicated that forests, woodland, and parkland had the highest total and above ground carbon content demonstrating potential for carbon sequestration. For example, total system carbon in the Senegal River valley was 115 ton ha^{-1} in the forest zone and only 18 ton ha^{-1} when the land was under cultivation. The cultivated systems have reduced carbon contents due to reduced tree cover and increased mineralization due to surface disturbance.

Generally, it seemed that SOC reduced with sampling depth at all sites used for the study. The continuously and conventionally tilled plots were among the plots with the lowest soil pH probably because of mining of exchangeable cations by growing crops in continuously tilled plots. Generally, soil pH increased with soil depth in most of the sites studied. Mineralogy, surface charge characteristics and precipitation of amorphous Fe and Al oxides on clay mineral surfaces define the capacity of clay minerals to adsorb and potentially protect SOC [21].

Results of this study also indicate that although site 3 was conventionally tilled and cultivated for 12 consecutive years, it stored up to 7025 gC m^{-2}. This may be because crop residues were always left in the field after harvesting but more importantly because it is a floodplain. It is likely that soil materials including C may have been transported from other places and deposited there. However, for site 9 (fallowed for 2 years, conservation till + fertilizer + poultry droppings and planted with maize) carbon stock was 3604 gC m^{-2} which was higher than the C values for plots 10 and 2 (conventionally tilled, continuously cropped plots) by up to 23%.

The quantity of carbon stored in the natural forest was greater than that of the artificial forest by 5% probably because of greater diversity of plant species found at the natural forests and to a lesser extent because the natural forests are older than the artificial forests. However, [21] and [22] have shown that both natural and artificial forest attain steady state conditions after several years and thereafter only slight changes in SOC content are possible unless extraneous factors like climatic shifts occur.

These results show that conventional tillage reduces soil carbon stocks when compared to other management practices. However, the amounts and rates of carbon sequestration vary according to natural factors such as climate (temperature and rainfall) and soil physical characteristics (soil texture, clay mineralogy, and soil depth) as well as agricultural management practices.

1.4 RESULTS

1.4.1 Soil Properties of the Study Sites

Results of the study (Table 2) indicate low, medium, and high coefficients of variability among soil properties at the different sites studied. There was a low coefficient of variability (6–9%) in bulk density and soil pH in $CaCl_2$ at the different soil depths studied, whereas silt + clay content and percent sand content showed medium variability (20–30%). The highest SOC content (3.07%) was found in site No. 6 (natural undisturbed forest) (Table 3), whereas lowest SOC was observed in site No. 10 (conventionally tilled, continuously cropped plot (CT-CC) (0.81%) and site No. 2 (CT-CC Plot) (0.83%) SOC. Lowest SOC levels were found in sites 2, 5, and 10 (CT-CC plots) with SOC range of between 0.59–0.83%. Ratings by Landon [16] in the study area show 1.16% SOC or lower to be low, whereas SOC values ≥1.74% and above are regarded as high. Sites 2 and 10 as shown in Table 1, were conventionally tilled and continuously cropped soils.

TABLE 2 Selected soil properties of the study sites.

S/N	Bulk	Density	(Mgm^{-3})	Organic	Carbon	($MgKg^{-1}$)	Total	N	($MgKg^{-1}$)	pH	($Cacl_2$)		Clay + silt	$MgKg^{-1}$		Sand	$MgKg^{-1}$	
	0–5	5–15	15–30	0–5	5–15	15–30	0–5	5–15	15–30	0–5	5–15	15–30	0–5	5–15	15–30	0–5	5–15	15–30
1.	1.47	1.47	1.49	1.25	1.22	1.20	0.126	0.111	0.101	5.0	5.0	5.3	56	56	58	44	44	42
2.	1.52	1.55	1.56	0.83	0.80	0.61	0.070	0.041	0.036	4.9	5.2	5.2	36	34	34	64	66	66
3.	1.30	1.32	1.37	2.31	2.13	1.98	0.130	0.126	0.120	5.3	5.3	5.5	26	28	28	74	72	72
4.	1.36	1.37	1.39	2.79	2.70	2.44	0.24	0.211	0.18	4.8	4.6	4.6	38	38	40	62	62	60
5.	1.29	1.31	1.32	0.63	0.62	0.59	0.09	0.09	0.07	4.3	4.3	4.6	34	35	38	66	65	62
6.	1.31	1.32	1.35	3.07	2.90	2.72	1.95	1.90	1.61	4.5	4.7	4.7	36	37	37	64	63	63
7.	1.23	1.25	1.28	2.32	2.46	2.66	0.20	0.20	0.15	4.9	4.6	4.5	37	37	39	63	63	61
8.	1.60	1.62	1.66	1.63	1.57	1.52	0.27	0.22	0.21	5.2	5.4	5.4	28	30	30	72	70	70
9.	1.44	1.45	1.48	1.06	1.00	0.94	0.085	0.082	0.080	5.04	5.0	4.5	35	37	37	65	63	63
10.	1.49	1.51	1.54	0.81	0.72	0.70	0.042	0.040	0.040	4.4	4.5	4.5	66	68	69	34	32	31
CV (%)	8.6	8.5	8.39	53.3	54.1	55.7	178	186	184	6.9	7.67	8.00	31.5	30.9	42.00	29.3	20.6	21.5

Anikwe, *Carbon Balance and Management* 2010; 5:5.

TABLE 3 Total quantity of SOC (gC m^{-2}) stored at the 0–30 cm soil layer of the study soils.

Site Number	0–5 (Mean + SEM)	Soil Depth 5–15 (Mean + SEM)	15–30 (Mean + SEM)	Total
1.	918.8 ± 12	1793.4 ± 2	1788 ± 44	4500.2
2.	630.8 ± 28	1240 ± 15	951.6 ± 23	2822.4
3.	1501.5 ± 34	2811.6 ± 33	2712.6 ± 20	7025.7
4.	1897.2 ± 26	3699 ± 18	3391.6 ± 26	8987.8
5.	387.5 ± 18	812.2 ± 15	778.8 ± 31	1978.5
6.	2010.9 ± 15	3828 ± 30	3672 ± 15	9510.9
7.	1426.8 ± 24	3075 ± 42	3404.8 ± 56	7906.6
8.	1308 ± 33	2551.5 ± 46	2523.2 ± 13	6382.7
9.	763.2 ± 12	1450 ± 21	1391.2 ± 17	3604.4
10.	603.5 ± 30	1087.2 ± 34	1078 ± 20	2768.7
CV (%)	65.3	54.5	56.3	55.1%

Anikwe, *Carbon Balance and Management* 2010; 5:5.

The highest total N content of the soils ranged from 0.29–1.95 Mg kg^{-1}. These were found at sites 6, 7, and 8. These plots were either artificially planted forests or natural undisturbed forests (Table 1), whereas sites 2, 5, and 10 had low N content, and correspond to plots that were conventionally and continuously tilled. Results show slight differences in pH values for the different soils studied. However, sites 5 and 10, which were continuously and conventionally tilled plots, were among the plots with the lowest soil pH.

Results of the study show that there were differences in total quantity of carbon sequestered in the different land utilization types in the study area (Table 3). These differences were confirmed by the high coefficient of variation (55%) between the SOC content of the different land use types.

The highest quantities of SOC were stored in sites 6, 4, and 7 with 9510.9, 8987.8, and 7906.6 gC m^{-2} in the 0–30 cm soil layers, respectively (Table 3). These sites correspond to natural undisturbed forest, artificial forest, and artificial grassland, respectively. Only slight differences in carbon stock (absolute difference between maximum and minimum value: 1604 gC m^{-2}) were found between the three land uses with the highest carbon stocks and that may be either because of differences in plant biodiversity, differences in bulk densities of the soils studied or slight differences in local climatic regimes.

The lowest carbon stocks in the 10 study locations were found in sites 5, 2, and 10. These have SOC stocks of 1978.5, 2822.4, and 2768.7 gC m^{-2} in their 0–30 cm soil layer, respectively. These plots correspond to conventionally tilled and continuously cropped plots. When compared to the sites with the highest carbon stocks (forest and grassland land use types), results show 71% depletion in carbon stock in the conventionally tilled, and continuously cropped plots. More specifically, the quantity of carbon sequestered in site 4 (planted forest) was 8989.8 gC m^{-2}. This was higher than that stored in an adjacent cultivated site (site 5) by as much as 78% (Table 3). Assuming that this forest reached a steady state condition (balanced input and output of SOC), it took 15 years of continuous cultivation and conventional tillage to lose 78% of its carbon stock built over the years.

Results show that at site 8 (Abakaliki, Artificial *Gmelina arborea* forest with alleys cultivated with food crops), the quantity of carbon sequestered was 6382.7 gC m^{-2} at 0–30 cm soil depth. This quantity was higher than the carbon stock found in site 1 (another Abakaliki plot, conventionally tilled and continuously cropped by 30%). In contrast, only a slight difference (5%) in total carbon stock was found between site 6 (natural undisturbed forest and site 4 (artificial forest).

1.5 CONCLUSION

The results of this study have shown that different management systems impact on the ability of the soil to sequester carbon. In tropical hot climates as those found in the study area, natural undisturbed forests, artificial forests, and grasslands store between 7906–9510 gC m^{-2} within the first 0–30 cm soil layer, whereas cultivated and continuously cropped lands sequester about 1978–3604 gC m^{-2} depending on the management system adopted. In other words, the large-scale conversion of forests to croplands in the southeastern Nigeria may lead to 50–75% loss in the regional soil carbon stock.

KEYWORDS

- **Agroecosystems**
- **Clean development mechanism**
- **Global warming**
- **Macro-Kjeldahl method**
- **Soil management**
- **Soil organic carbon**

REFERENCES

1. Soil Science Society of America. *Carbon sequestration in soils Position of the Soil Science Society of America*. SSSA Ad Hoc Committee S893 Report, (2001).
2. Tieszen, L. L. Carbon sequestration in semi-arid and sub-humid Africa. U.S. Geological Survey, EROS Data Center, Sioux Falls, South Dakota (2000). Retrieved from http://edcintl.cr.usgs.gov/ip.
3. Huggins, D. R., Buyanvsky, G. A., Wagner, G. H., Brown, J. R., Darmody, R. G., Peck, T. R., Lesoing, G. W., Vanotti, M. B., and Bundy, L. G. Soil organic C in the tall grass prairie-derived region of the Corn Belt: effect of long-term management. *Soil and Tillage Research*, **47**, 227–242 (1998).
4. Schlesinger, W. H. Carbon sequestration in soils. *Nature*, **284**, 2095–2096 (1999).

5. Schlesinger, W. H. and Lichter, J. Limited carbon storage in soils and litter of experimental forest plots under increased atmospheric CO_2. *Nature*, **411**, 466–469 (2001).
6. Dick, W. A., Belvins, R. L., Frye, W. W., Peters, S. E., Christenson, D. R., Pirece, F. J., and Vitosh, M. L. Impact of agricultural management practices on C sequestration in forest-derived soils of the eastern Corn Belt. *Soil and Tillage Research*, **47**, 243–252 (1998).
7. West, T. O. and Post, W. M. Soil organic carbon sequestration rates by tillage and crop rotation . A global data analysis. *Soil Science Society Am Journal*, **66**, 1930–1946 (2002).
8. Raji, B. A. and Ogunwole, J. O. Potential of soil carbon sequestration under various land use in the sub-humid and semi-arid savanna of Nigeria, Lessons from long-term experiments. *International Journal of Soil Science*, **1**, 33–43 (2006).
9. Dumanski, J., Desjardins, R. L., Tarnocai, C., Monreal, D., Gregorich, E. G., Kirkwood, V., and Campbell, C. A. Possibilities for future carbon sequestration in Canadian agriculture in relation to land use changes. *Climate Change*, **40**, 81–103 (1998).
10. Gebhart, D. L., Johnson, H. B., Mayenx, H. S., and Polley, H. W. The CRP increases soil organic carbon. *Journal of Soil and Water Conservation*, **49**, 488–492 (1994).
11. Ramussen, P. E., Albrecht, S. L., and Smiley, R. W. Soil C and N changes under tillage and cropping systems in semi-arid Pacific Northwest agriculture. *Soil and Tillage Research*, **47**, 205–213 (1998).
12. Studdert, G. A. and Echeverría, H. E. Crop rotations and nitrogen fertilization to manage soil organic carbon dynamics. *Soil Science Society Am Journal*, **64**, 1496–1503 (2000).
13. Natural Resource Conservation Service. Soil survey staff: *keys to soil taxonomy* 9th ed. United States Department of Agriculture, Bethesda, Maryland (2003).
14. Campbell, C. A., Janzen, H. H., Paustian, K., Gregorich, E. G., Sherrod, L., Liang, B. C., and Zentner, R. P. Carbon storage in soils of the North American Great Plains: effect of cropping frequency. *Agron. Journal*, **97**, 349–363 (2005).
15. Ellert, B. H., Janzen, H. H., and McConkey, B. G. Measuring and comparing soil carbon storage. *Assessment methods for soil carbon*. CRC Press, Boca Raton, Florida pp. 131–146 (2001).
16. J. R. Landon (Ed.). *Booker Tropical Soil Manual a handbook for soil survey and agricultural land evaluation in the tropics and subtropics*. John Wiley and Sons, New York (1991).
17. Anikwe, M. A. N., Obi, M. E., and Agbim, N. N. Effect of crop and soil management practices soil compactibility in maize and groundnut plots in a Paleustult in southeastern Nigeria. *Plant and Soils*, **253**, 457–465 (2003).
18. Krull, E., Baldock, J., and Skjemstad, J. *Soil texture effects on decomposition and soil carbon storage*. In NEE Workshop Proceedings, CRC for Greenhouse Accounting, CSIRO Land and Water Australia, (April 18–20, 2001).
19. Lal, R. Soil carbon dynamic in cropland and rangeland. *Environmental Pollution*, **116**, 353–362 (2002).
20. Bationo, A., Kihara, J., Vanlauwe, B., Waswa, B., and Kimetu, J. Soil organic carbon dynamics, functions, and management in West African agroecosystems. *Agricultural Systems*, **94**, 13–25 (2007).
21. Cole, C. V., Flach, K., Lee, J., Sauerbeck, D., and Stewart, B. Agricultural sources and sinks of carbon. *Water, Soil, and Air Pollution*, **70**, 111–122 (1993).
22. Jastraw, J. D., Boutton, T. W., and Miller, M. M. Carbon dynamics of aggregate-associated organic matter estimated by carbon-13 natural abundance. *Soil Science Society of America Journal*, **60**, 801–807 (1996).
23. Unamma, R. P. A., Odurukwe, S. O., Okereke, H. E., Ene, L. S. O., and Okoli, O. O. *Farming Systems in Nigeria Report of the benchmark survey of the eastern agricultural zone of Nigeria*. NRCRI Umudike, Umuahia, Nigeria (1985).
24. Mulla, D. J. and McBratrey, A. B. Soil spatial variability. *Handbook of Soil Science*. CRC Books, New York (2000).
25. Doran, J. W. and Mielke, L. N. A rapid, low-cost method for determination of soil bulk density. *Soil Science Society of America Journal*, **48**, 717–719 (1984).

26. Nelson, D. W. and Sommers, L. E. Total carbon, organic carbon, and organic matter. In *Methods of soil analysis, Part 3: Chemical methods*. Soil Science Society of America Book Series 5. D. L. Sparks, A. L. Page, P. A. Helmke, R. H. Loeppert, P. N. Soltanpour, M. A. Tabatabai, C. T. Johnson, and M. E. Sumner. (Eds.). SSSA, Madison, Wisconsin, pp. 1961–1010 (1996).
27. Bremner, J. M. Total nitrogen. In *Methods of soil analysis, Part 3: Chemical methods*. Soil Science Society of America Book Series 5. D. L. Sparks, A. L. Page, P. A. Helmke, R. H. Loeppert, P. N. Soltanpour, M. A. Tabatabai, C. T. Johnson, and M. E. Sumner. (Eds.). SSSA, Madison, Wisconsin, pp. 1085–1122 (1996).
28. Gee, G. W. and Orr, D. Particle-size analysis. In *Methods of soil analysis, Part 4 Physical methods*. Soil Science Society of America Book Series 5. J. H Dane and G. C Topp. (Eds). SSSA, Madison, Wisconsin, pp. 255–293 (2002).

2 Soil and Tillage Management in the Northwest Great Plains

Zhengxi Tan, Shuguang Liu, Zhengpeng Li, and Thomas R Loveland

CONTENTS

2.1 INTRODUCTION

Tillage practices greatly affect carbon (C) stocks in agricultural soils. Quantification of the impacts of tillage on C stocks at a regional scale has been challenging because of the spatial heterogeneity of soil, climate, and management conditions. We evaluated the effects of tillage management on the dynamics of soil organic carbon (SOC) in croplands of the Northwest Great Plains ecoregion of the United States using the general ensemble biogeochemical modeling system (GEMS). Tillage management scenarios included actual tillage management (ATM), conventional tillage (CT), and no-till (NT).

The model simulations show that the average amount of C (kg C $ha^{-1}yr^{-1}$) released from croplands between 1972 and 2000 was 246 with ATM, 261 with CT, and 210 with NT. The reduction in the rate of C emissions with conversion of CT to NT at the ecoregion scale is much smaller than those reported at plot scale and simulated for other regions. Results indicate that the response of SOC to tillage practices depends significantly on baseline SOC levels: the conversion of CT to NT had less influence on SOC stocks in soils having lower baseline SOC levels but would lead to higher potentials to mitigate C release from soils having higher baseline SOC levels.

For assessing the potential of agricultural soils to mitigate C emissions with conservation tillage practices, it is critical to consider both the crop rotations being used at a local scale and the composition of all cropping systems at a regional scale.

Many studies have identified the potential of soils cultivated with conservation practices (e.g., NT) to sequester large amounts of C [1, 2]. It is estimated that conservation tillage practices across the United States may drive large-scale sequestration on the order of 24–40 Tg C yr^{-1} (Tg: teragram; 1 Tg = 1012 g), and that additional C sequestration of 25–63 Tg C yr^{-1} can be achieved through other modifications to traditional agricultural practices [3]. In regard to the C credit scenario established by the Kyoto Protocol, it is widely suggested that conversion of CT to NT can help to support the profitability of C credits for farmers. The uncertainties of these sequestration scenarios, however, depend on SOC monitoring and/or models [2].

Recently, eddy-covariance measurements have been used to evaluate the contribution of NT practice to C dynamics in corn *(Zea mays* L.) and soybean (*Glycine max* (L.) Merr.) rotation ecosystems at regional and national scales [1, 2]. However, the relationships between net ecosystem exchange (NEE) and either terrestrial C storage or actual SOC stocks are still poorly understood, owing to the uncertainty of the redistribution of biomass in farming products beyond a given C accounting region. For example, based on the assessment of NEE over 6 years, Hollinger et al. [2] estimated C stocks for corn/soybean rotation ecosystems in the North Central Region of the United States, and observed a C sink under NT at the local scale which, however, is not necessarily true on a regional scale. They attributed this discrepancy to regional consumption of grain combined with C emissions associated with agricultural practices.

The temporal variability in SOC stock is indicative of the response of ecosystems to changes in climate, land use/land cover, and land management. Because the dynamics of SOC directly impact the availability of nutrients and moisture to all kinds of living organisms, changes in SOC stock can transform the structure and functions of ecosystems, and may also result in ecosystem feedbacks on climate [4]. The GEMS [5] has been used by Tan et al. [6] to simulate the terrestrial C dynamics in the Northwest Great Plains between 1972 and 2000. Results show that C sources of croplands and the SOC balances across the ecoregion depend on the proportion of cropped area to grassland. This study, however, did not take into account the contribution of land management to SOC dynamics.

How well we predict future atmospheric CO_2 dynamics and their response to anthropogenic CO_2 emissions depends on understanding of the extent to which the rising atmospheric CO_2 concentration can be offset by terrestrial ecosystems through conservation agricultural practices. Therefore, we need to evaluate the impacts of change trends in land use/land cover and land management on terrestrial C source-sink relationships associated with specific management practices.

Sequestering C in cultivated soils managed with NT is being advocated as a way to assist in meeting the demands of an international C credit system [1]. The potential of cropland with such conservation management to mitigate CO_2 emissions from mixed grass-crop ecosystems at a regional scale needs to be evaluated. In this study, we simulated SOC dynamics within the top 20 cm of soil in the cropped areas of the Northwest Great Plains between 1972 and 2000 with three management scenarios: The ATM, all cropland managed with CT, and all cropland managed with NT.

2.2 METHODS

2.2.1 Study Area

The study area is the Northwest Great Plains ecoregion, with a land area of 338,718 km^2 (Figure 1). The average annual precipitation from 1972 to 2000 was 400 mm (wetter in the eastern portion of the ecoregion), and the average annual temperature was 7.2°C (warmer in the southern portion of the ecoregion) [6]. The mean annual temperature was 0.67°C higher between 1986 and 2000 than between 1972 and 1985.

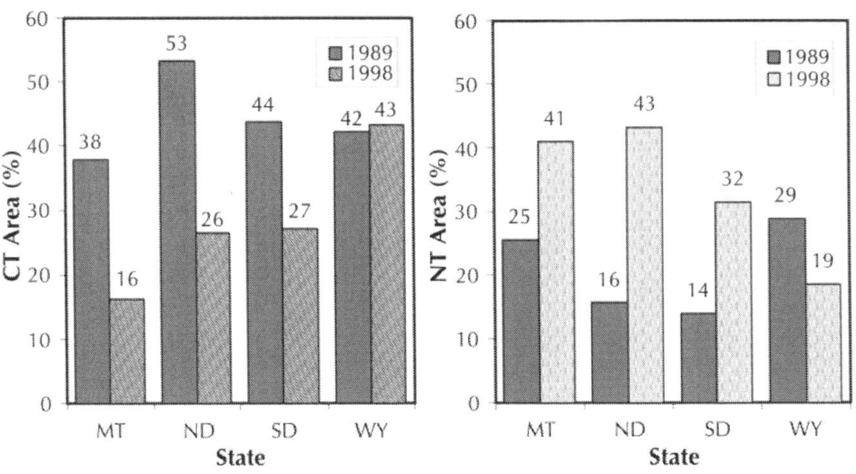

FIGURE 1 The study area and locations of sample blocks (After Tan et al. [7]).

As illustrated in Figure 1, the land cover in the ecoregion consisted of 75% mixed grasses, 17% cropland, and 8% other land covers, on average, from 1972 to 2000 [6]. Agriculture, however, has been the primary land use transforming this grassland dominated ecosystem. Cumulative change in land cover during the period accounted for about 10% of all land area, but most of these changes were directly related to conversions between cropland and grassland that resulted from the implementation of the conservation reserve program (CRP) [6]. Cropped areas were mainly located on level ground where soils were fertile and devoted to row crops, small grains, and fallow. Major crop types included spring wheat (*Triticum aestivum*), corn (*Zea mays L.*), soybean (*Glycine max (L.) Merr.*), and alfalfa (*Medicago sativa*). The proportion of all cropland to the total land area varied from state to state, ranging from 5% in Wyoming to 46% in North Dakota.

2.2.2 Sampling Framework

The sampling protocol proposed by Loveland et al. [8] was introduced to identify spatial variation of land use/land cover change in the conterminous United States using Omernik's 84 Level III Ecoregions [9] as the sampling framework. For the Northwest Great Plains ecoregion, forty sample blocks of 10 × 10 km each were randomly selected (Figure 1) to identify changes with a precision of 1% at 85% confidence level [8]. Changes were detected based on five dates (1973, 1980, 1986, 1992, and 2000) of landsat multispectral scanner (MSS) and thematic mapper (TM) imagery data that were analyzed at a cell size of 60 × 60 m for MSS images and 30 × 30 m for TM images.

2.2.3 Modeling System

The GEMS [5] was used to simulate soil C dynamics in this study. The GEMS is a modeling system developed for a better integration of spatially explicit time series land use and land cover change data with well-established ecosystem biogeochemical models (e.g., CENTURY). The GEMS has been used to simulate C dynamics in vegetation and soil for diverse ecosystems, especially in the Northwest Great Plains [6, 7]. As described by Liu et al. [5] and Tan et al. [7], the GEMS consists of three major components: single or multiple encapsulated ecosystem biogeochemical models, an automated stochastic parameterization system (AMPS), and an input/output processor (IOP). The AMPS includes two major interdependent parts: the data search and retrieval algorithms and the data processing mechanisms. The first part searches for and retrieves relevant information from various databases according to the keys provided by a joint frequency distribution (JFD) table. The data processing mechanisms downscale the aggregated information at the map unit level to the field scale using a Monte Carlo approach. Once the data are assimilated, they are injected into the modeling processes through the IOP which updates the default input files with the assimilated data. The values of selected output variables are also written by the IOP to a set of output files after each model execution. The JFD grids are first created from soil maps, a time series of land cover images, and climate themes at a cell size of 60 × 60 m. The CENTURY model [23] was selected as the underlying ecosystem biogeochemical model in GEMS for this study because it has solid modules for simulating C dynamics at the ecosystem level and has been widely applied to various ecosystems worldwide.

2.2.4 Input Data for Model

The spatial simulation unit of GEMS is a JFD case. The JFD case contains single or multiple, homogeneous, connected or isolated land pixels that represent a unique combination of values from the geographic information system (GIS) layers [5]. The data for model input primarily consisted of climatic regimes, land use/land cover change, soil inventory, management data, nitrogen (N) deposition map, and administrative districts. The land use/land cover data were described. Climatic data consisted of annual precipitation and maximum and minimum temperature records from 1973 through 2000, which were converted to 30 m pixel GIS layers from the CRU TS 2.0 datasets [24]. Soil characteristics within each sample block were taken from the US State Soil Geographic Data Base (STATSGO) [25] for initializing the soil components of GEMS. Tillage management data were derived from the conservation technology information center (CTIC) [17].

The GEMS automates the processes of downscaling forest ages from the United States department of agriculture (USDA) forest inventory and analysis data (FIA), crop compositions from the agricultural census, and grass cover distribution and temporal changes from the remotely sensed imagery interpretation. A Monte Carlo method was used to assign each JFD a set of specific soil property values such as layer depth, soil organic matter content, soil water holding capacity, and clay and sand percentages. Based on the definitions set by CENTURY, we partitioned the SOC stock into different pools at the beginning of each simulation using a retrospective SOC initialization algorithm: the slow SOC pool was defined from the net primary productivity (NPP) for each land cover type and soil inventory data; the difference between the total SOC and the slow pool was then used to initialize the passive SOC pool; and the active SOC pool was set at about 2% of the total SOC storage [7].

2.2.5 Ensemble Simulations

The GEMS generates site-level inputs with a Monte Carlo approach from regional data sets. Any single simulation of a JFD case is unique combination of randomly picked forest age, crop species, and soil properties from regional level datasets, so that the output of a single simulation run of a JFD might be biased. Therefore, ensemble simulations of each JFD were executed to incorporate the variability of inputs and to average uncertainties of simulation results. In general, the averages of ensemble simulations become more stable when increasing the times of run. We made 20 repeat runs for each JFD case in this study, which reduced the relative error to about 2%. The averaged JFD output from the 20 runs was then aggregated at sample block scale, and the simulation uncertainty was evaluated on both sample block and the ecoregion scales. In this study, values of selected output variables were written to a set of output files after each model execution, and then aggregated at four spatial levels: pixel (60 × 60 m) → land use category → sample block (10 × 10 km) → ecoregion.

2.2.6 Tillage Management Data and ATM Scenario

Tillage management data were collected from the CTIC [24] for areas of annual CT, reduced tillage (RT) and CT practices (including mulch, ridge, and NT) and 8 crop types at a county scale for the period from 1989 to 1998. The areas for each combination of cropping system and tillage management were documented for individual counties at a two-year interval. For each combination of county, year, and crop type, the area extents were determined for the three tillage options.

The ATM information for model simulations was based on the CTIC data with an assumption that all planted area was managed with CT in 1972 and then converted to NT and RT until 1989 at a pace similar to that estimated from the CTIC data for the period between 1989 and 1998. The areas under ATM for all cropped lands are presented in Figure 2. The probability for each tillage management option for the years after 1998 was assumed to be the same as that in 1998. Tillage information for each crop was read in from the CTIC dataset to determine the three proportions of the total planted area for CT, RT, and NT. For a certain type of crop (e.g., corn), the fraction of each tillage type in 1992 was assigned as 0.5, 0.3, and 0.2 to planted areas managed with CT, RT, and NT, respectively.

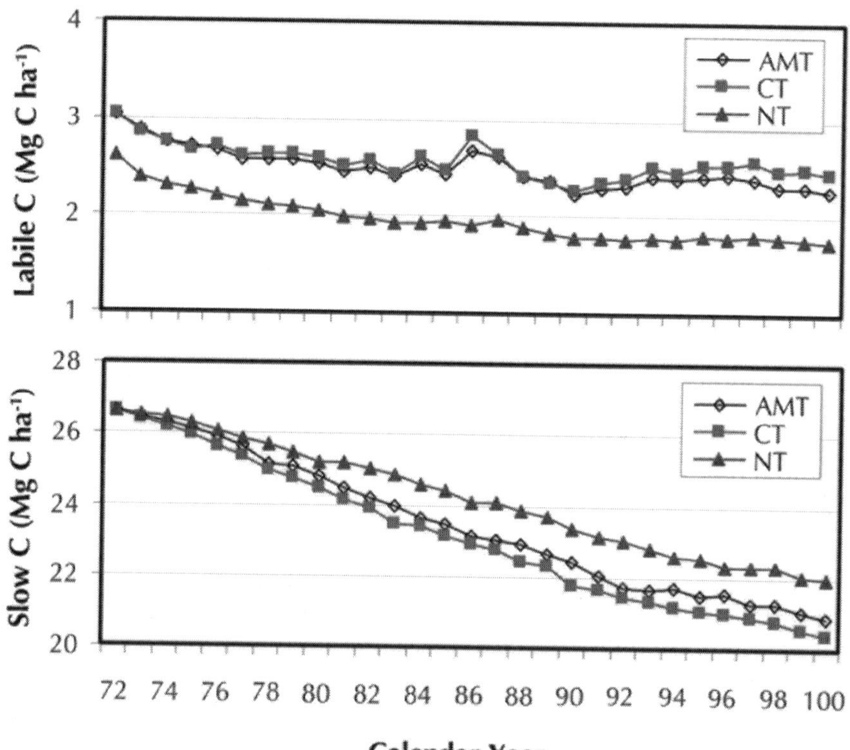

FIGURE 2 Areal percentage of each type of tillage management for the planted area of the Northwest Great Plains ecoregion (CT, NT, and RT).

2.2.7 CT and NT Scenarios for Model Simulations

To evaluate the effects of different tillage options on SOC dynamics across the ecoregion, we defined two extreme tillage scenarios for model simulations: (1) complete CT, which assumed that all planted areas were managed with CT since 1972, and (2) complete NT, in which all planted areas were assumed to be managed with NT since 1972.

2.3 DISCUSSION

The results indicate that the reduction in the rate of C emissions with conversion of CT to NT at the ecoregion scale is much smaller than those reported at the plot scale, and also much smaller than simulated values for other regions. For example, based on field scale measurements of C flux between the atmosphere and the corn/soybean ecosystem in the United States for the years 1997–2002, Bernacchi et al. [1] demonstrate that current corn/soybean agricultural practices release more C than is removed from the atmosphere with 10% of the cropland being in continuous NT agriculture, but that this ecosystem can become a C sink estimated at a rate of 300 kg C ha^{-1} yr^{-1} with NT management implemented over a larger area. In fact, the responses of the SOC stock in the upper soil layer to tillage practices not only demonstrate considerable spatial variation, but also depend on cropping systems at local scales [11], and are even influenced by sampling protocols [12].

Many studies have shown that NT does not necessarily lead to C sequestration within the upper 20 cm depth of soil; NT can result in either C sources or sinks, depending on cropping systems [13, 14]. Although relatively high reduction rates of C release have been reported for continuous corn and corn/soybean cropping systems, other cropping systems such as continuous soybean, cotton production, and wheat/summer fallow rotation are usually reported as C sources [13, 15, 16]. As indicated by the data presented in Table 1, for either the complete CT or the complete NT scenario in this study, two-thirds of the planted area was in small grain production dominated by wheat, with only about 12% for corn and 8% for soybeans. The management data (named as CTIC data afterwards) derived from the CTIC [17] show that the fallow area was equivalent to 18% of the total planted area. This composition of cropping systems would determine the extent to which the SOC stock had been influenced by tillage practices during the study period.

Huggins et al. [11] assessed the effects of crop sequence and tillage on SOC stocks using the natural 13C abundance of corn (*Zea mays L.*) and soybeans (*Glycine max (L.), Merr.*). Soil samples were collected after 14 years under each treatment for SOC quantification. They observed the influence of crop sequence on SOC (0–45 cm depth) that occurred when tillage was reduced with chisel plow and NT. Results show that there was 15% more SOC in continuous corn than in continuous soybean, but all tillage treatments within continuous soybean systems showed little influence on SOC. Peterson et al. [18] conducted CT-NT paired experiments on wheat-dominated cropping systems in Mandan, North Dakota and found a small annual increase rate (about 0.25%) of the SOC pool with NT, which is very close to result averaged at the ecoregion scale.

Halvorson et al. [16] documented that the SOC stock did not increase during 12 years under a spring wheat/fallow system with NT in North Dakota. Compared with CT, the NT tended to lose more SOC at a rate of 50 kg ha^{-1} yr^{-1} in the upper 15 cm depth under a spring wheat/fallow system, which seems to support the conclusion that soil C in the surface layer can be quickly lost to the atmosphere by increasing summer fallow practice [19]. A similar result was also reported by Campbell et al. [20] in Canada. Halvorson et al. [16] suggested that conversion from crop/fallow to more intensive cropping systems with NT is needed in order to have a positive impact on reducing CO_2 emissions from croplands in the Northern Great Plains ecoregion.

Based on metadata analyses with a large number of point observations from published works, Manley et al. [21] concluded that NT is less effective for sequestering C on the Prairies than in other regions (e.g., in the southern United States and the Corn Belt), and also less effective with wheat than with other crops. Generally, NT for either a continuous corn system or a corn/soybean rotation system leads to C sinks and more net C gain comes with a longer duration of NT [2], which, however, depends on baseline SOC contents [10] and tends to level off as the soil becomes saturated [22]. Unfortunately, this study could not define the saturation level for the whole study area because we do not enough long-term management data and necessary field observations to drive model simulations. Furthermore, the saturation level varies greatly not only with specific soil but also with other many factors.

Using historical county-level land use data for the 19th and 20th centuries to drive an ecosystem model, Parton et al. [15] conducted four case studies within the Great Plains of the United States that were used to represent different agro-ecosystems. Model results show that cultivation of grassland results in large losses of SOC and an increase in soil nitrogen mineralization for the first 20 to 30 years of cultivation, followed by small SOC loss and nitrogen mineralization after 50 years cultivation. Their simulation results also indicate that the irrigated cotton production would lead to a net C source whereas the irrigated corn and alfalfa cropping systems would result in a C sink in the central and northern Great Plains.

The estimate of the limited reduction in the rate of SOC release with conservation tillage management across the Northwest Great Plains could be attributed to the wheat/fallow-dominated crop rotation and the composition of all cropping systems being practiced in this ecoregion.

2.4 RESULTS

2.4.1 Croplands and Tillage Management History

The major historical changes in land use/land cover within the ecoregion were directly related to conversions between cropland and grassland. In 1972, the average percentages of cropland and grassland were 17% and 75%, respectively, and changed to 15% and 77%, respectively, in 2000. For the study area, the average percentage of cropland from 1972 to 2000 was 41, 21, 9, and 5% in North Dakota, Montana, South Dakota, and Wyoming, respectively.

Average cropped area between 1989 and 1998 was about 4.06 million ha within the ecoregion. As indicated in Figure 3, the CT area was 45% in 1989 and decreased to 24% in 1998. During the same period, the area of NT increased from 19 to 38%, and the RT area changed slightly from 36 to 38%. But the change rate of each tillage-managed area varied from state to state (see Figure 3). For example, from 1989 to 1998, the NT area increased by 27% in North Dakota whereas it declined by 10% in Wyoming.

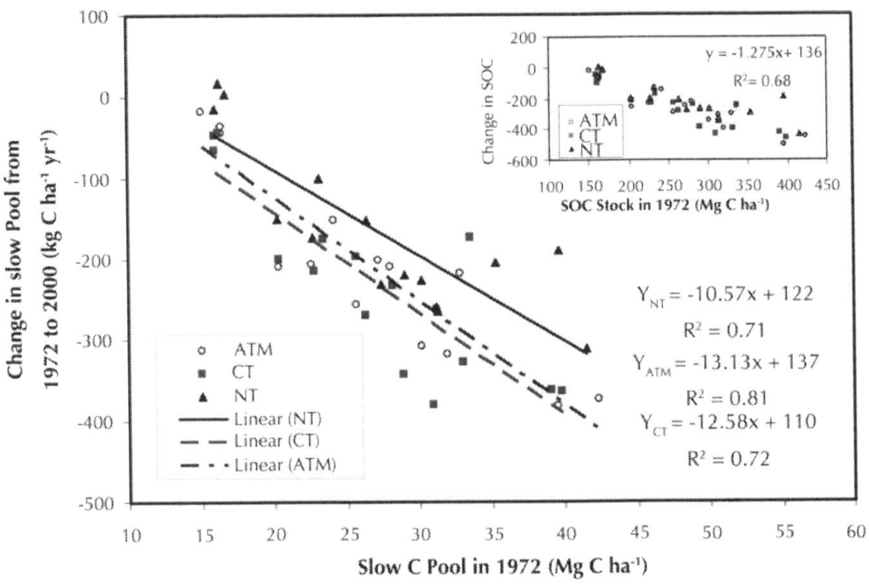

FIGURE 3 Changes in CT and NT areas for each state between 1989 and 1998 in the Northwest Great Plains ecoregion.

2.4.2 Cropping Systems and Associations with Tillage Management

During the 1990s, there was little variation in the area proportions of major cropping systems in the ecoregion, which consisted of 67% small grain crops (predominated by wheat), 12% corn, 8% soybean, and 14% others (Table 1). The areal proportion of each cropping system managed with NT, RT, and CT was 30%, 35%, and 35%, respectively, and not significantly different over time despite large differences among the four states.

TABLE 1 The percentage of individual cropping systems in the total planted area under each tillage management at the ecoregion scale.

Cropping system	NT	RT	CT	Total
Corn	4.2	3.7	3.6	11.5
Small grain	20.4	24.8	21.4	66.6
Soybean	2.9	2.4	3.1	8.4
Other	2.7	4.5	6.9	14.0
Total planted[a]	29.9	35.2	34.9	100.0

NT: no-till; RT: reduced tillage; CT: conventional tillage.
[a] excluding fallow area.

2.4.3 Changes in SOC Pools with Tillage Management Scenarios

For sample blocks with cropped area percentage greater than 10%, total SOC stock within the top 20 cm depth of soil generally tended to decrease with cultivation time, but the reduction in SOC stock was smaller under NT (210 ± 33 kg C $ha^{-1}yr^{-1}$) than under both CT (261 ± 36 kg C $ha^{-1}yr^{-1}$) and ATM (246 ± 38 kg C $ha^{-1}yr^{-1}$). The average reduction in the rate of C release with conversion of CT to NT was about 51 kg C $ha^{-1}yr^{-1}$ during the study period. No significant difference between ATM and CT was observed. The reduction in the rate of C release, however, was correlated with the baseline SOC levels. For example, sample block 04 in North Dakota had a high baseline SOC stock (66 Mg C ha^{-1}) and was simulated to have a reduction in the rate of C release at 104 kg C $ha^{-1}yr^{-1}$ with NT compared with CT. In contrast, sample block 02 in Montana had a low baseline SOC stock (34.5 Mg C ha^{-1}) and was simulated to show a reduction in the rate of C emission of 56 kg C $ha^{-1}yr^{-1}$ with NT in comparison with CT. Simulation results also demonstrate that, by the year 2000, sample block 04 remained a C source at the rate of 330 kg C $ha^{-1}yr^{-1}$, whereas sample block 02 turned into a C sink at the rate of 9 kg C $ha^{-1}yr^{-1}$.

Simulation results indicate that the changes in total SOC stock with conversion of CT to NT were predominantly a result of changes in both the labile and slow C pools. Figure 4 shows that there was a relatively consistent reduction (about 0.7 Mg C ha^{-1}) in the labile SOC pool under NT in comparison with CT (and ATM); whereas the slow pool increased by 1.6 Mg C ha^{-1} by the year 2000. In other words, the conversion of CT to NT would reduce the slow C emissions at the rate of 57 kg C $ha^{-1}yr^{-1}$ in the study area.

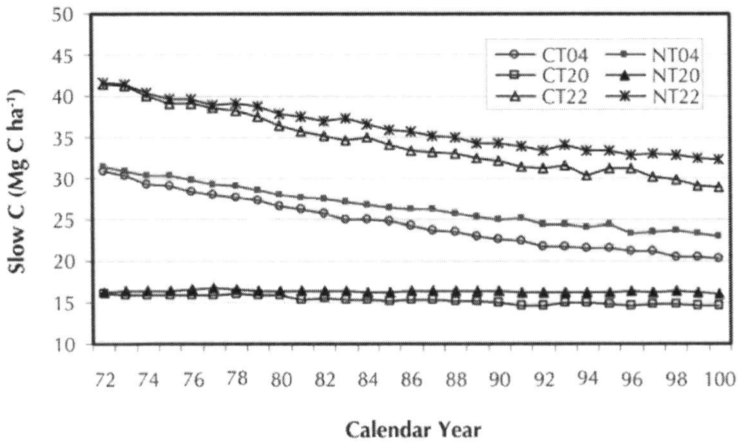

FIGURE 4 Temporal trends in soil labile C and slow C pools in association with CT, NT, and ATM within sample blocks where cropped area was greater than 10% of the block area in the Northwest Great Plains ecoregion.

2.4.4 Change Rate of SOC Pools in Relation to Baseline SOC Levels

The data in Figure 5 indicate that crop production, regardless of tillage practices, tends to remove SOC from the top soil layer, even though there is less loss under

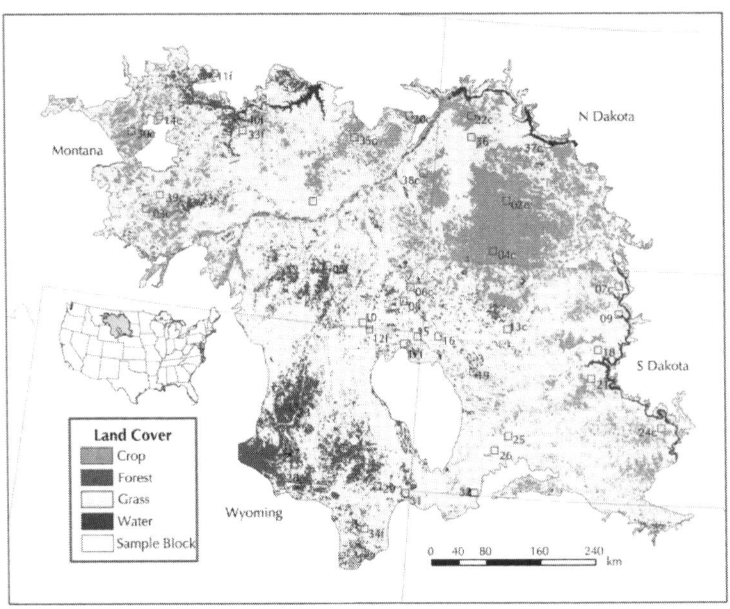

FIGURE 5 Magnitude of the change in the slow C pool in the top 20 cm depth of soil from 1972 to 2000 for three tillage scenarios in relation to the baseline slow pool in 1972 for the sample blocks in which the cropped area percentage was greater than 10%. The inset graph illustrates the relation of the change in total SOC stock to the baseline.

NT than under CT. Responses of total SOC stock (especially of the slow C pool) to tillage management, however, depend significantly on baseline SOC levels. As illustrated in Figure 6, soils with higher baseline SOC content tend to lose more C following crop cultivation. Conversely, the conversion of CT to NT has less influence on SOC stocks for soils having lower baseline SOC levels, but would lead to higher potentials to mitigate the C release from soils having higher baseline SOC content. This is in agreement with the conclusion of Tan et al. [10] for the east central United States.

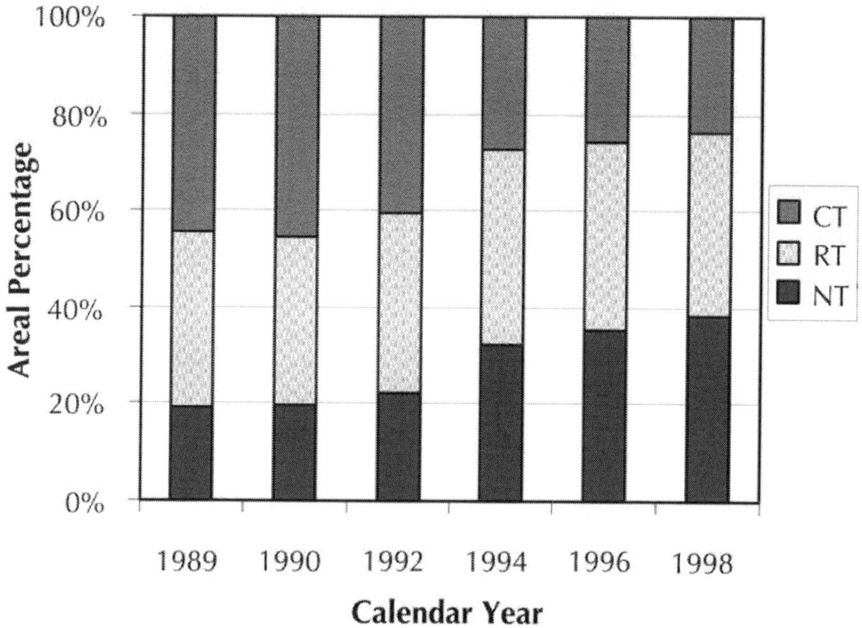

FIGURE 6 Temporal trends in slow SOC pools under CT and NT in association with various baselines SOC levels in 1972 at the sample block scale within the Northwest Great Plains ecoregion (The numbers 04, 20, and 22 refer to the sample block IDs. These three blocks are located in North Dakota, Montana, and North Dakota, respectively).

2.5 CONCLUSION

The simulated reduction in the rate of C release with conversion of CT to NT in the agricultural soils across the Northwest Great Plains ecoregion between 1972 and 2000 is much smaller than those reported from plot scale studies and also smaller than simulated values for other regions in general. However, similar estimates are reported by other investigators for the crop rotations and composition of cropping systems similar to those in study area. The changes in total

SOC stock were predominantly a result of the dynamics of the slow C pool at the study's time span. We suggest that the responses of total SOC to tillage management scenarios depend significantly on the baseline SOC level. Soils with higher SOC levels tend to have higher potentials to reduce C emissions with conservation tillage practices, but the dominance of the wheat/fallow crop rotation and the composition of all cropping systems could be the primary cause for the limited efficiency of NT for mitigating C emissions as simulated for the Northwest Great Plains ecoregion.

KEYWORDS

- **Conventional tillage**
- **Geographic information system**
- **Reduced tillage**
- **Soil organic carbon**
- **Tillage management data**

AUTHORS' CONTRIBUTIONS

Zhengxi Tan designed this study, prepared model input, analyzed results, and drafted the manuscript. Shuguang Liu supervised this research and did model debug as the developer of GEMS. Zhengpeng Li was in charge of programming for incorporating tillage management data into model simulations and model debug. Thomas R Loveland provided the time series land use/land cover change trends data and comments on the manuscript formulation. All authors read and approved the final manuscript.

ACKNOWLEDGMENT

This research was part of the U.S. Carbon Trends Program that was supported by the Geographic Analysis and Monitoring Program of the U.S. Geological Survey (USGS). We thank Norman Bliss for his constructive suggestions and technical review. We also thank Bruce Wylie, John Hutchinson, and four anonymous reviewers for their valuable comments. Work of Zhengxi Tan, Shuguang Liu, and Zhengpeng Li performed under USGS contract 03CRCN0001.

REFERENCES

1. Bernacchi, C. J., Hollinger, S. E., and Tilden, M. The conversion of the corn/soybean ecosystem to no-till agriculture may result in a carbon sink. *Global Change Biology*, **11**, 1867–1872 (2005).
2. Hollinger, S. E., Bernacchi, C. J., and Meyers, T. P. Carbon budget of mature no-till ecosystem in North Central Region of the United States. *Agricultural and Forest Meteorology*, **130**, 59–69 (2005).
3. R. Lal, J. M. Kimble, R. F Follett, and C. V. Cole (Eds.). *The potential of us cropland to sequester carbon and mitigate the greenhouse effect*. Lewis Publishers, New York, **128** (1999).

4. Hayden, B. P. Ecosystem feedbacks on climate at the landscape scale. Philosophical Transactions of the Royal Society of London, Series B, *Biological Sciences*, **353**, 5–18 (1998).
5. Liu, S., Loveland, T. R., and Kurtz, T. R. Contemporary Carbon Dynamics in Terrestrial Ecosystems in the Southeastern Plains of the United States. *Environmental Management*, **33**, 442–456 (2004).
6. Tan, Z., Liu, S., Johnston, C. A., Loveland, T. R., Tieszen, L. L., Liu, J., and Kurtz, R. Soil organic carbon dynamics as related to land use history in the northwestern Great Plains. *Global Biogeochem Cycles*, **19** (2005).
7. Tan, Z., Liu, S., Johnston, C. A., Liu, J., and Tieszen, L. L. Analysis of ecosystem controls on soil carbon source-sink relationships in the northwest Great Plains. *Global Biogeochem Cycles*, **20**, GB4012 (2006).
8. Loveland, T. R., Sohl, T. L., Stehman, S. V., Gallant, A. L., Sayler, K. L., and Napton, D. E. A strategy for estimating the rates of recent United States land-cover changes. *Photogrammetric Engineering and Remote Sensing*, **68**, 1091–1099 (2002).
9. Omernik, J. M. Ecoregions of the conterminous United States. *Annals of the Association of American Geographers*, **77**, 118–125 (1987).
10. Tan, Z., Lal, R., and Liu, S. Using experimental and geospatial data to estimate regional carbon sequestration potential under no-till practice. *Soil Science*, **171**, 950–959 (2006).
11. Huggins, D. R, Allmaras, R. R, Clapp, C. E, Lamb, J. A, and Randall, G. W. Corn-Soybean Sequence and Tillage Effects on Soil Carbon Dynamics and Storage. *Soil Science Soc Am J*, **71**, 145–154 (2007).
12. Dolan, M. S., Clapp, C. E., Allmaras, R. R., Baker, J. M., and Molina, J. Soil organic carbon and nitrogen in a Minnesota soil as related to nitrogen, tillage, residue, and nitrogen management. *Soil and Tillage Research*, **89**, 221–231 (2006).
13. Sainju, U. M., Lenssen, A. W., Caesar, T., and Waddell, J. T. Tillage and crop rotation effects on dry land soil and residue carbon and nitrogen. *Soil Sci Soc Am J*, **70**, 668–678 (2006).
14. West, T. O. and Post, W. M. Soil organic carbon sequestration rates by tillage and crop rotation: A global data analysis. *Soil Sci Soc Am J*, **66**, 1930–1946 (2002).
15. Parton, W. J., Gutmann, M., Williams, S. A., Easter, M., and Ranchhod, V. The Ecological Impact of Historical Land Use Patterns in the Great Plains: A Methodological Assessment. *Ecological Applications*, **15**, 1915–1928 (2005).
16. Halvorson, A. D., Wienhold, B. J., and Black, A. L. Tillage, nitrogen, and cropping system effects on soil carbon sequestration. *Soil Sci Soc Am J*, **66**, 906–912 (2002).
17. Conservation Technology Information Center (CTIC). National crop residue management survey data. Retrieved from http://www.conservationinformation.com.
18. Peterson, G. A., Halvorson, A. D., Havlin, J. L., Jones, O. R., Lyon, D. J., and Tanaka, D. L. Reduced tillage and increasing cropping intensity in the great plains conserves soil C. *Soil Till Res*, **47**, 207–218 (1998).
19. Curtin, D., Wang, H., Selles, F., McConkey, B. G., and Campbell, C. A. Tillage effects on carbon fluxes in continuous wheat and fallow-wheat rotations. *Soil Sci Soc Am J*, **64**, 2080–2086 (2000).
20. Campbell, C. A., Zentner, R. P., Liang, B. C., Roloff, G., Gregorich, E. C., and Blomert, B. Organic C accumulation in soil over 30 years in semiarid southwestern Saskatchewan—Effect of crop rotations and fertilizers. *Canadian Journal of Soil Science*, **80**, 179–192 (2000).
21. Manley, J., Van Kooten, G. C., Moeltner, K., and Johnson, D. W. Creating carbon offsets in agriculture through no-till cultivation: a meta-analysis of costs and carbon benefits. Department of Economics University of Victoria, *In Climatic Change*, **68**, 41–65 (2005).
22. Antle, J. A. and McCarl, B. A. The economics of carbon sequestration in agricultural soils. In *the international year book of environmental and resource economics 2002/2003*. T. Tietenberg and H. Folmer (Eds.). Edward Elgar ,Cheltenham, UK, pp. 278–310 (2002).
23. Parton, W. J., Scurlock, M. O., Ojima, D. S., Gilmanov, T. G., Scholes, R. J., Schimel, D. S., Kirchner, T., Menau, J C., Seastedt, T., Moya, E., Kamnalrut, A., and Kinyamario, J. I. Observa-

tions and modeling of biomass and soil organic matter dynamics for the grassland biome worldwide. *Global Biogeochem Cycles*, **7**, 785809 (1993).
24. Climatic Research Unit: CRU TS 2.0 datasets. Retrieved from http://www.cru.uea.ac.uk/cru/data/hrg.htm
25. USDA-NRCS: State Soil Geographic (STATSGO) Data Base: Data use information. Miscellaneous Publication No. 1492 (1994).

3 Biochar: Carbon Mitigation from the Ground Up

David J. Tenenbaum

CONTENTS

3.1 INTRODUCTION

As multibillion-dollar projects intended to sequester carbon dioxide (CO_2) in deep geologic storage continue to seek financial support, the fertile black soils in the Amazon basin suggest a cheaper, lower tech route toward the same destination. Scattered patches of dark, charcoal-rich soil known as terra preta (Portuguese for "black earth") are the inspiration for an international effort to explore how burying biomass derived charcoal, or "biochar," could boost soil fertility and transfer a sizeable amount from the atmosphere into safe storage of CO_2 topsoil. Although burial of biochar is just beginning to be tested in long-term, field-scale trials, and studies of Amazonian terra preta show that charcoal can lock up carbon in the soil for centuries and improve soil fertility.

Charcoal is made by heating wood or other organic material with a limited supply of oxygen (a process termed "pyrolysis"). The products of the pyrolysis process vary by the raw material used, burning time, and temperature, but in principle, volatile hydrocarbons and most of the oxygen and hydrogen in the biomass are burned or driven off, leaving carbon-enriched black solids with a structure that resists chemical and microbial degradation. Christoph Steiner, a research scientist at the University of Georgia, says the difference between charcoal and biochar lies primarily in the end use. "Charcoal is a fuel, and biochar has a nonfuel use that makes carbon sequestration feasible". "Otherwise there is no difference between charcoal carbon and biochar carbon."

Charcoal is traditionally made by burning wood in pits or temporary structures, but modern pyrolysis equipment greatly reduces the air pollution associated with this practice. Gases emitted from pyrolysis can be captured to generate valuable products instead of being released as smoke. Some of the by-products can be condensed into "bio-oil," a liquid that can be upgraded to fuels including biodiesel and synthesis gas. A portion of the noncondensable fraction is burned to heat the pyrolysis chamber and the rest can provide heat or fuel an electric generator.

Pyrolysis equipment now being developed at several public and private institutions typically operates at 350–700°C. In Golden, Colorado, Biochar Engineering Corporation is building portable $50,000 pyrolyzers that researchers will use to produce 1–2 tons of biochar per week. Company CEO Jim Fournier says the firm is planning larger units that could be trucked into position. Biomass is expensive to transport, he says, so pyrolysis units located near the source of the biomass are preferable to larger, centrally located facilities, even when the units reach commercial scale.

3.2 BIOCHAR BUILDS BETTER SOIL

Spanish conquistador Francisco de Orellana reported seeing large cities on the Amazon River in 1541, but how had such large populations raised their food on the poor Amazonian soils? Low in organic matter and poor at retaining plant nutrients which makes fertilization inefficient—these soils are quickly depleted by annual cropping. The answer lay in the incorporation of charcoal into soils, a custom still practiced by millions of people worldwide, according to Steiner. This practice allowed continuous cultivation of the same Amazonian fields and thereby supported the establishment of cities.

Researchers who have tested the impact of biochar on soil fertility say that much of the benefit may derive from biochar's vast surface area and complex pore structure which is hospitable to the bacteria and fungi that plants need to absorb nutrients from the soil. Steiner says, "We believe that the structure of charcoal provides a secure habitat for *microbiota*, which is very important for crop production." Steiner and coauthors noted in the 2003 book *Amazonian Dark Earths* that the charcoal-mediated enhancement of soil caused a 280–400% increase in plant uptake of nitrogen.

The contrast between charcoal enriched soil and typical Amazonian soil is still obvious, says Clark Erickson, a professor of anthropology at the University of Pennsylvania. Terra preta stands out, because the surrounding soils in general are poor, red, oxidized, and so rich in iron and alluminum that they sometimes are actually toxic to plants. Today, patches of terra preta are often used as gardens.

Anna Roosevelt, a professor of anthropology at the University of Illinois at Chicago, believes terra preta was created accidentally through the accumulation of garbage. The dark soil is full of human cultural traces such as house foundations, hearths, cemeteries, food remains, and artifacts, along with charcoal. In contrast, Erickson said that the Amazonian peoples knew exactly what they were doing when they developed this rich soil. As evidence, Erickson says "All humans produce and toss out garbage, but the terra preta phenomenon is limited to a few world regions."

Recent studies show that although biochar alone does not boost crop productivity, biochar plus compost or conventional fertilizers make a big difference. In the February 2007 issue of Plant and Soil, Steiner, along with Cornell University soil scientist Johannes Lehmann and colleagues demonstrated that use of biochar plus chemical amendments (nitrogen–phosphorus–potassium fertilizer and lime) on average doubled grain yield over four harvests compared with the use of fertilizer alone. In research presented at the April 2008 235th national meeting of the American Chemical Society, Mingxin Guo, an assistant professor of agriculture at Delaware State University, found that biochar plus chemical fertilizer increased growth of winter wheat and several vegetables by 25–50% compared with chemical fertilization alone.

"Depending on the sources, biochar may supply certain amounts of phosphorus and potassium to crops but will supply little nitrogen," says Guo. "On the other hand, biochar promotes growth of beneficial microbes and helps retain phosphorus and potassium in soil, improving crop utilization efficiency of the nutrients. Nevertheless, biochar fertilization may initially require more nitrogen from external sources since decomposition of biochar carbon will consume available nitrogen in soil." With the decrease in phosphorus fertilization and increase in nutrient retention, biochar should have positive effects on reducing nutrient runoff losses, according to Guo, who adds, "Since biochar fertilization enhances soil aeration and beneficial microbial activity, it will also inhibit soilborne pathogens but not aboveground pests."

However, not all biochar performs the same. The importance of biochar's variable chemical composition was illustrated in studies by Goro Uehara, a professor of soil science at the University of Hawaii, who grew plants both with and without biochar made from macadamia nutshells. Goro Uehara says, "As we added more [biochar], the plants got sicker and sicker." Uehara's colleague, University of Hawaii extension specialist Jonathan Deenik, says that when they repeated the experiment with a more highly carbonized version of the nutshell biochar, which contained lower levels of volatile compounds, "preliminary results in a greenhouse study showed that low-volatility [biochar] supplemented with fertilizer outperformed fertilizer alone by 60%, in a statistically significant difference." This research was presented at the October 2008 annual meeting of the soil science society of America.

3.3 BANKING CARBON

Researchers have come to realize the use of biochar also has phenomenal potential for sequestering carbon in a warming world. The soil already holds 3.3 times as much carbon as the atmosphere, according to a proposal Steiner wrote for submission by the United Nations Convention to Combat Desertification (UNCCD) to the Ad Hoc Working Group on Long-term Cooperative Action at the 1–10 December 2008 United Nations climate conference in Poznan, Poland. However, Steiner wrote, many soils have the capacity to hold probably several hundred billions of tons more.

Plants remove CO_2 from the atmosphere through photosynthesis, and then store the carbon in their tissues. CO_2 is released back into the atmosphere after plant tissues decay or are burned or consumed and the CO_2 is then mineralized. If plant materials are transformed into charcoal, however, the carbon is permanently fixed in a solid form—evidence from Amazonia, where terra preta remains black and productive after several thousand years, suggests that biochar is highly stable. On average, half the biochar carbon is recalcitrant and would persistently remain in soil, according to Guo.

Carbon can also be stored in soil as crop residues or humus (a more stable material formed in soil from decaying organic matter). But soil chemist Jim Amonette of the Department of Energy's Pacific Northwest National Laboratory points out that crop residues usually oxidize into CO_2 and are released into the atmosphere within a couple of years and the lifetime of carbon in humus is typically less than 25 years.

The calculations for potential carbon storage can be estimated downward from the amount of atmospheric carbon that photosynthesis removes from the air each year; using figures from the intergovernmental panel on climate change, Amonette estimates that number at 61.5 billion metric tons. Amonette said that the best estimates are presented in four scenarios for carbon storage calculated by the nonprofit International Biochar Initiative (IBI), a consortium of scientists and others who advocate for research/development and commercialization of biochar technology. The IBI's "moderate" scenario assumed that 2.1% of the annual total photosynthesized carbon would be available for conversion to biochar containing 40% of the carbon in the original biomass, and that incorporating this charcoal in the soil would remove half a billion metric tons of carbon from the atmosphere annually. Because the heat and chemical energy released during pyrolysis could replace energy derived from fossil fuels, the IBI calculates the total benefit would be equivalent to removing about 1.2 billion metric tons of carbon from the atmosphere each year. That would offset 29% of today's net rise in atmospheric carbon which is estimated at 4.1 billion metric tons, according to the Energy Information Administration.

It is these large numbers combined with the simplicity of the technology that has attracted a broad range of supporters. At Michigan Technological University, for example, undergraduate Amanda Taylor says she is "interested in changing the world" by sequestering carbon through biochar. Under the guidance of department of humanities instructor Michael Moore, Taylor, and fellow students established a research group to study the production and use of biochar as well as how terra preta might fit into a framework of community and global sustainability. Among other projects, the students made their own biochar in a 55 gallon drum and found that positioning the drum horizontally produced the best burn.

The numbers are entirely theoretical at this point and any effort to project the impact of biochar on the global carbon cycle is necessarily speculative, says Lehmann. "These estimates are at best probing the theoretical potential as a means of highlighting the need to fully explore any practical potential, and these potentials need to be looked at from environmental, social, and technological viewpoints. The reason we have no true prediction of the potential is because biochar has not been fully tested at the scale that it needs to be implemented at to achieve these predictions."

Still, Steiner stresses that other large-scale carbon storage possibilities also face uncertainties. "Forests only capture carbon as long as they grow, and the duration of sequestration depends very much on what happens afterward," Steiner stresses says. "If the trees are used for toilet paper, the capture time is very short." Soilborne charcoal, in contrast, is more stable, he says: "The risk of losing the carbon is very small—it cannot burn or be wiped out by disease, like a forest."

As a carbon mitigation strategy, most biochar advocates believe biochar should be made only from plant waste, not from trees or plants grown on plantations. "The charcoal should not come from cutting down the rainforest and growing eucalyptus," says Amonette.

3.4 A STEP TOWARD LEGITIMACY

Biochar took a step toward legitimacy at the December Poznan conference, when the UNCCD placed it in consideration for negotiations for use as a mitigation strategy during the second Kyoto Protocol commitment period, which begins in 2013. Under the cap-and-trade strategy that forms the backbone of the Kyoto Protocol, businesses can buy certified emission reduction (CER) credits to offset their emissions of greenhouse gases. If biochar is recognized as a mitigation technology under the Kyoto clean development mechanism, people who implement this technology could sell CER credits. The market price of credits would depend on supply and demand a high enough price could help promote the adoption of the biochar process.

The possibility that the United Nations will give its stamp of approval to biochar as a climate mitigation strategy means the ancient innovation may finally undergo large-scale testing. "The interest is growing extremely fast, but it took many years to receive the attention," says Steiner. "Biochar for carbon sequestration does not have strong financial support compared to carbon capture and storage through geological sequestration. However, biochar is much more realistic for carbon capture."

"For now," he says, "think the biggest hope and advantage is carbon sequestration and the ability to address sustainable land use, food and renewable energy production, and carbon sequestration in a complementary way—not a competing way."

KEYWORDS

- **Banking carbon**
- **Biochar**
- **Carbon**
- **Charcoal**

SUGGESTED READINGS

Amonette J.; Lehmann J.; Joseph S. Terrestrial carbon sequestration with biochar: a preliminary assessment of its global potential. Eos Trans AGU 2007, 88(52 Fall Meeting Suppl): Abstract U42A-06.

BioEnergy Lists: Terra Preta (Biochar) website. Available: http://terrapreta.bioenergylists.org/

Glazer B.; Woods W.I., editors. Amazonian dark earths: explorations in space and time. New York: Springer Verlag, 2004.

International Biochar Initiative. How much carbon can biochar systems offset—and when? Available: http://www.biochar-international.org/images/final_carbon.pdf.

Lehmann J.; Joseph S. Biochar for environmental management: science and technology. London: Earthscan, 2009.

Woods W.I.; Teixeira W.G.; Lehmann J.; Steiner C. Amazonian dark earths: Wim Sombroek's vision. New York: Springer Verlag, 2009.

4 Tropical Litterfall and CO_2 in the Atmosphere

Emma J. Sayer, Jennifer S. Powers, and Edmund V. J. Tanner

CONTENTS

4.1 INTRODUCTION

The aboveground litter production in forests is likely to increase as a consequence of elevated atmospheric carbon dioxide (CO_2) concentrations, rising temperatures, and shifting rainfall patterns. As litterfall represents a major flux of carbon from vegetation to soil, changes in litter inputs are likely to have wide-reaching consequences for soil carbon dynamics. Such disturbances to the carbon balance may be particularly important in the tropics because tropical forests store almost 30% of the global soil carbon, making them a critical component of the global carbon cycle nevertheless, the effects of increasing aboveground litter production on belowground carbon dynamics are poorly understood. We used long-term, large-scale monthly litter removal and addition treatments in a lowland tropical forest to assess the consequences of increased litterfall on belowground CO_2 production. Over the second to the fifth year of treatments, litter addition increased soil respiration more than litter removal decreased it; soil respiration was on average 20% lower in the litter removal and 43% higher in the litter addition treatment compared to the controls but litter addition did not change microbial biomass. We predicted a 9% increase in soil respiration in the litter addition plots, based on the 20% decrease in the litter removal plots and an 11% reduction due to lower fine root biomass in the litter addition plots. The 43% measured increase in soil respiration was therefore 34% higher than predicted and it is possible that this "extra" CO_2 was a result of priming effects, that is stimulation of the decomposition of older soil organic matter by the addition of fresh organic matter. The results show that increases in aboveground litter production as a result of global change have the potential to cause considerable losses of soil carbon to the atmosphere in tropical forests.

The changes in litter quantity as a consequence of global climate change are increasingly likely; recent FACE experiments have shown that litterfall increases with elevated atmospheric CO_2 concentrations [1-5] and predicted changes in rainfall distribution patterns [6] and temperature [7] may also affect litterfall by altering leafing phenology. As litterfall represents a major pathway for carbon and nutrients between vegetation and soil it seems likely that changes in aboveground litter production will have consequences for belowground processes. However, despite the increasing recognition that research on terrestrial ecosystem dynamics needs a combined aboveground-belowground approach [8], the potential impact of changes in litterfall on belowground carbon dynamics has been largely ignored [9].

Tropical forests are a critical component of the global carbon cycle as they store 20–25% of the global terrestrial carbon [10, 11]. Ongoing debates about whether tropical forests are a source or sink for atmospheric carbon have led to increased interest in the belowground components of their carbon cycle [12] because they also contribute almost 30% to global soil carbon storage [13]. Soil respiration from root and heterotrophic respiration alone releases approximately 80 Pg of carbon into the atmosphere per year to which tropical and subtropical forests contribute more than any other biome [14]. Recent studies have investigated the direct effects of elevated CO_2 [15], rising temperature [16, 17], and fertilizer [18-20] on soil carbon cycling but we know very little about how soil respiration will be affected by the predicted changes in aboveground production caused by global climate change. We believe that an increase in aboveground litterfall may have a large impact on belowground carbon and nutrient cycling, as annual litterfall is closely correlated with soil respiration on a global scale [21, 22], and the amount of litter on the forest floor also affects soil nutrient status, soil water content, soil temperature, and pH [9], all of which can influence soil respiration rates. To investigate this, we conducted an experiment consisting of large-scale monthly litter removal (L-) and litter addition (L+) treatments in a lowland tropical forest. The results show that an increase in aboveground litterfall caused a disproportionate increase in soil respiration, reduced the amount of carbon allocated to fine root biomass and thus has the potential to cause substantial losses of carbon belowground.

4.2 MATERIALS AND METHODS

4.2.1 Site Description

The study was carried out as part of an ongoing long-term litter manipulation experiment to investigate the importance of litterfall in the carbon dynamics and nutrient cycling of tropical forests. The forest under study is an old-growth moist lowland tropical forest, located on the Gigante Peninsula (9°06'N, 79°54'W) of the Barro Colorado Nature Monument in Panama, Central America. The soil is an oxisol with a pH of 4.5–5.0, with low "available" phosphorus concentration, but high base saturation and cation exchange capacity [28, 40, 41]. Nearby Barro Colorado Island (about 5 km from the study site) receives a mean annual rainfall of 2600 mm and has an average temperature of 27°C [42]. There is a strong dry season from January to April with a median rainfall of less than 100 mm per month [43]; almost 90% of the annual precipitation occurs during the rainy season. Fifteen plots (45 m × 45 m) were established within a 40 ha area (500 m×800 m) of old-growth forest in 2000. In 2001, all fifteen plots were trenched to a depth of 0.5 m in order to minimize lateral nutrient and water movement *via* the root/mycorrhizal network, the trenches were double lined with plastic and backfilled. Starting in January 2003, the litter (including branches ≤ 100 mm in diameter) in five plots was raked up once a month, resulting in low, but not entirely absent, litter standing crop (L- plots). The removed litter was immediately spread on five further plots, approximately doubling the monthly litterfall (L+ plots), five plots were left as controls (CT plots). The assignment of treatments was made on a stratified random basis, stratified by total litterfall per plot in 2001 that is, the three plots with highest litterfall were randomly assigned to treatments, then the next three and so on.

4.2.2 Soil Respiration

In October 2002, four measurement sites were established in each of the fifteen plots, collars made of PVC pipe of 108 mm inner diameter and 44 mm depth were placed 12.5 m into the plot, measured from the centers of each of the four sides of the plot. The collars were sunk into the soil to 10 mm depth and anchored using small plastic tent pegs which were attached to the collars by cable binders and sunk diagonally into the ground to avoid channeling water into the soil under the collars. The collars were left undisturbed throughout the experiment. Soil respiration from the mineral soil was measured in the collars in February, March, May, June, September, and October of 2003, and April, May, and September of 2004 using an infra-red gas analyzer (IRGA) Li-6400 with an Li-6400-9 soil chamber attachment (LI-COR, Lincoln, USA). The ambient CO_2 level was determined for each site individually and measurements started at 5 ppm below the ambient CO_2 level. Three measurements were taken over each collar at each time and the values were averaged to give one value per collar per time.

In September 2005, new measurement sites were set up adjacent to those established in 2002. The PVC collars of 200 mm diameter and 120 mm depth were sunk into the ground to 20 mm and anchored with tent pegs as described. The collars were set up two months before starting measurements and were left undisturbed throughout the experiment. Soil respiration was measured over the new collars in November and December 2005, January, March, April, May, and June 2006, and May, June and July 2007 using the Li-8100 soil CO_2 flux system (LI-COR, Lincoln, USA). The ambient

CO_2 level was determined for each site automatically and one measurement of 2 min duration was taken over each collar. As main research aim was to determine whether the amount of litter on the forest floor affects respiration from the mineral soil, leaf litter was removed from the collars prior to all measurements, great care was taken not to disturb the underlying mineral soil and the litter was replaced once the measurements had been completed.

All measurements were made during 1–3 days each month between 8.00 hr and 14.00 hr. If measurements could not be completed in one day, they were made over consecutive days whenever possible, but no measurements were taken during or immediately following heavy rainfall. When measurements were taken over several days, an equal number of plots per treatment was measured each day; the plot means (four collars per plot) were used for statistical analysis of treatment effects.

The annual respiration rates were estimated by obtaining a daily mean each for the rainy and dry seasons from the data collected, multiplying the daily mean by the average number of days in each season (135 days for the dry season and 230 days for the rainy season), and then summing the obtained values.

4.2.3 Soil Temperature and Soil Water Content

Soil temperature was recorded during respiration measurements in 2003, 2004, 2006, and 2007 within 0.5 m of the collars using the IRGA's integrated soil temperature probe inserted to a depth of 100 mm. Volumetric soil water content was measured from 0–60 mm depth using a thetaprobe (Delta-T Devices, Cambridge, UK), which was calibrated to the soil type in the plots following the procedure described by Delta-T. Due to technical problems, volumetric soil water measurements were not made in 2004 or 2005 and only in April in 2006. Gravimetric soil water content was determined in May, June, and August 2004, and in January and June 2006 from four 20 mm diameter soil cores per plot, taken from 0–100 mm depth, volumetric soil water content was then calculated from the gravimetric measurements.

4.2.4 Fine Root and Microbial Biomass

The biomass of fine roots (≤ 2 mm diameter) from 0–100 mm depth in the mineral soil was determined in June and July 2004 from ten randomly located 51 mm diameter soil cores per plot [25], and in June 2006 from seven randomly located soil cores per plot, live and recently dead fine roots were carefully separated from the soil by washing in a 0.5 mm mesh sieve and then dried to constant weight in the oven at 70°C.

Total microbial biomass of the mineral soil was measured in August 2004 and June 2006 (during the rainy season). Four soil cores were taken from 0–100 mm depth at the four corners of the inner 20 m × 20 m in each plot using a 20 mm diameter punch-corer, the cores were bulked to give one sample per plot. Subsamples were taken to determine soil gravimetric water content and total microbial biomass was determined by the fumigation extraction method [44, 45]. Briefly, pairs of unfumigated and chloroform fumigated (exposed to chloroform for three days in the dark) soil samples were shaken in 0.5 M K_2SO_4 for 1 hr, filtered through Whatman No. 1 filter paper, and frozen until analyzed. Total organic carbon (TOC) and total nitrogen in the extracts were measured simultaneously on a TOC VCPH/CPN and nitrogen analyzer (Schimadzu,

Kyoto, Japan). Total microbial carbon and nitrogen were estimated as the difference between fumigated and unfumigated samples (expressed on an oven dry mass basis), divided by appropriate conversion factors [44, 45].

4.2.5 Statistical Analyses

Using mean values per plot (i.e., n = 5 for each treatment and control) differences among treatments in soil respiration rates, soil temperature, and soil moisture were investigated by separate repeated measures analysis of variance (ANOVAs) for each year. Fine root biomass and microbial biomass C and N were analyzed with one way ANOVAs for each year separately. Where treatment effects were found to be significant or marginally significant ($P < 0.07$), post hoc comparisons were made using Fisher's least significant difference (LSD) test. All analyses were carried out in Genstat 7.2 (VSN International Ltd., Hemel Hempstead, UK).

4.3 DISCUSSION

We attribute the lack of significant responses in soil respiration to the experimental treatments during the 1st year to a combination of two factors: firstly we started treatments during the dry season when decomposition is limited by the lack of moisture [23] and secondly, we did not include the litter layer in our measurements of soil respiration. Thus, we would not expect CO_2 efflux from the mineral soil to be affected by treatments until decomposition processes were sufficiently advanced to affect the input of carbon and nutrients to the mineral soil.

The 20% reduction in soil respiration observed in the litter removal treatment from the 2nd year of treatments until the end of the study is similar to the 28% decrease reported in plots in young regrowth forest in Brazil, after 1 year of litter removal where controls included the litter layer in CO_2 efflux measurements [24], but lower than the 51% decrease after 7 years of litter removal in lower montane forest in Puerto Rico [25]. We can attribute the decrease in study principally to a reduction in heterotrophic respiration due to the withdrawal of fresh substrate, as there were no differences in fine root biomass in the upper 100 mm of the mineral soil between CT and L- plots in 2004 or in 2006 (Figure 1). Furthermore, we found no significant differences in soil water content between treatments and the small ($\leq 0.5°C$) and inconsistent differences in soil temperature were unlikely to affect soil respiration.

We expected an average increase of 20% in soil respiration in the L + plots during the period from May 2004 to July 2007, as CO_2 efflux from the mineral soil decreased by this percentage in the L- plots. However, soil respiration in the L+ plots was on average 43% higher than the controls and therefore 23% higher than expected by the addition of litter alone. Furthermore, this increase was sustained from May 2004 until the end of the study in July 2007 (Figure 2). The increase in soil respiration in the L+ plots is considerable and greater than the effects of fertilization with 150 kg ha^{-1} yr^{-1} of phosphorus in a study in Costa Rica [20]. While fertilization treatments are thought to boost soil respiration by removing the nutrient limitation of decomposition processes [20, 26], and increasing microbial biomass [26], we found no increase in leaf litter decomposition rates in L+ treatments [27], and no changes in microbial biomass C or N (Figure 3(a),(b)). Furthermore, litter addition decreased fine root biomass in the mineral soil by 37% in 2004 [28] and 30% in 2006 (Figure 1). Fine roots contribute the bulk of root respiration [29, 30], root respiration is proportional to root biomass [31], and it typically makes up at least 30% of total soil respiration in the tropics [31-34]. Consequently the lower fine root biomass in the L+ plots would effectively reduce soil respiration by c. 11%. The expected increase in soil respiration in the L+ plots due to litter addition and reduced fine root biomass would therefore only be c. 9% relative to the controls. Thus, measured soil respiration was 34% higher than expected from the extra litter and reduced root biomass (Figure 4). This suggests that litter addition, besides increasing the amount of readily degradable carbon, may also cause substantial losses of CO_2 from the soil. It is likely that this extra CO_2 production is attributable to priming effects, the enhanced microbial decomposition of older, more recalcitrant soil organic matter by the addition of fresh organic matter [35, 36]. Strong pulses of soil respiration have been observed in lowland tropical rainforest during the dry-to-rainy season transition and were interpreted as a "natural priming effect" caused by large amounts of water soluble carbon leaching from the litter that had accumulated during the dry season [20]. This study shows that additional leaf litter has the potential to sustain priming effects throughout the year.

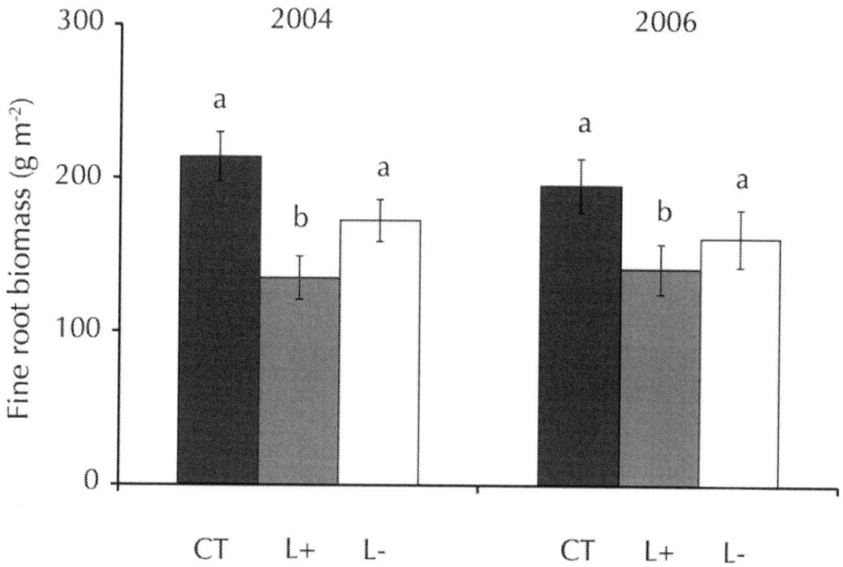

FIGURE 1 Fine root biomass in the mineral soil (0–100 mm) in litter manipulation treatments in tropical rainforest.

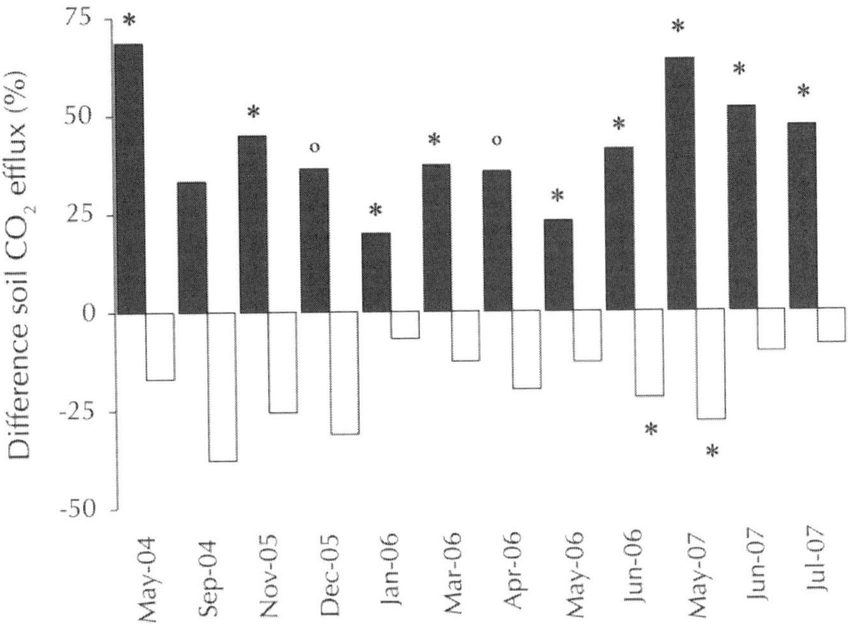

FIGURE 2 The differences in soil respiration between litter manipulation treatments and controls in tropical rainforest.

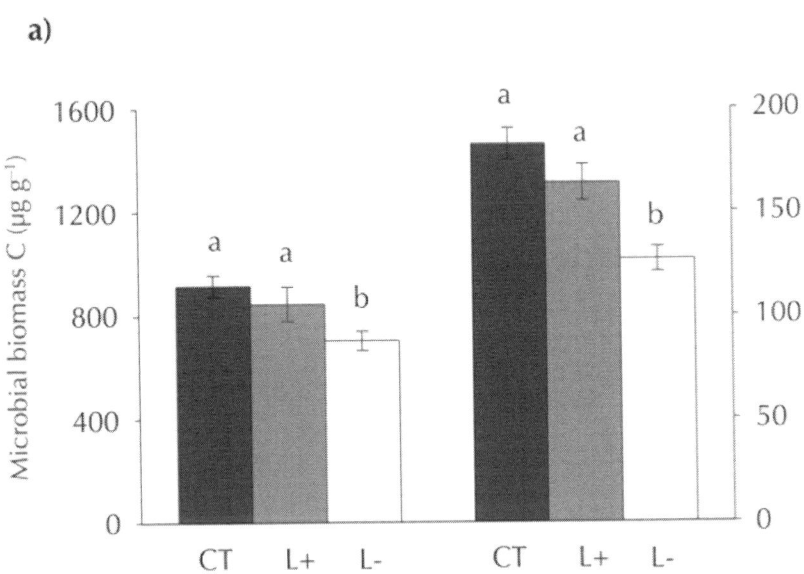

FIGURE 3 *(Continued)*

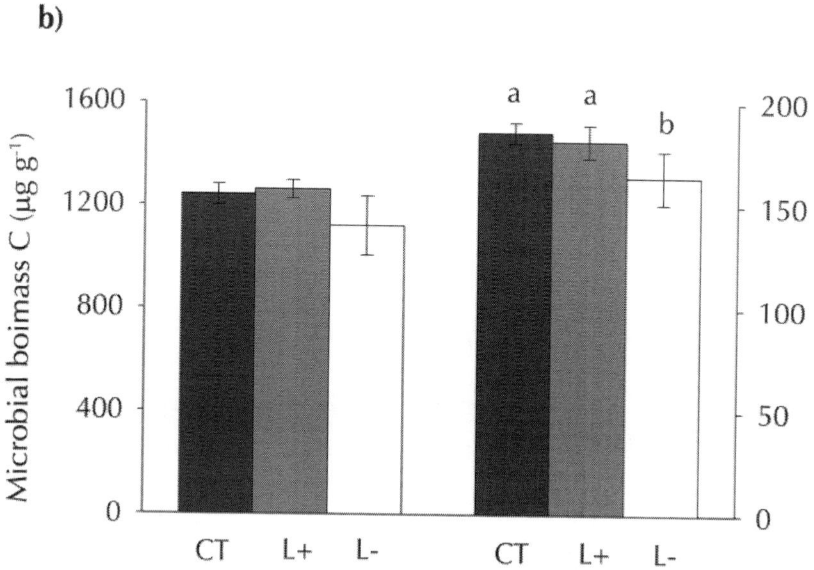

FIGURE 3 Microbial biomass in the mineral soil (0–100 mm) in litter manipulation treatments in tropical rainforest.

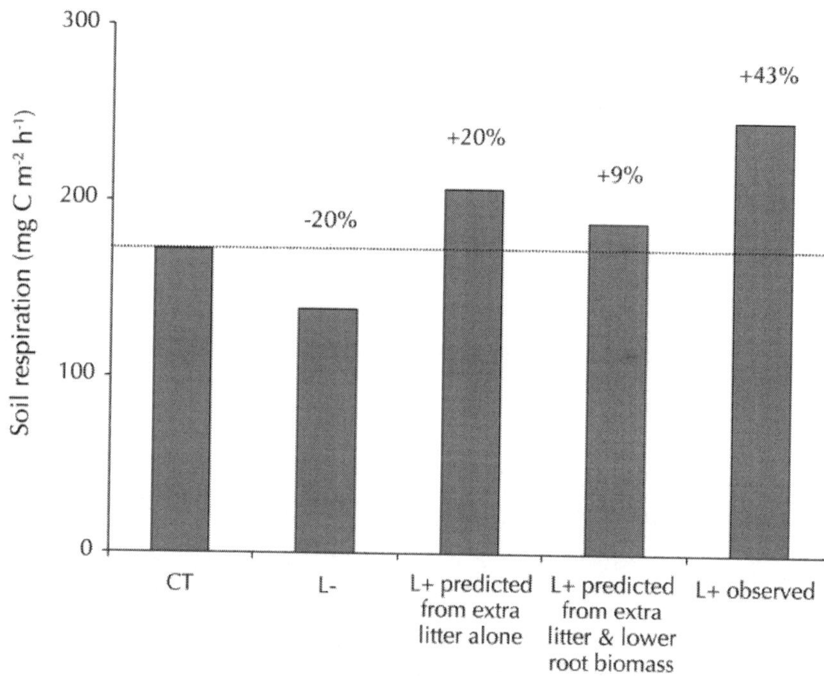

FIGURE 4 Comparison of predicted and observed soil respiration in litter manipulation plots in tropical rainforest.

The differences are expressed as percentages of the mean rate measured in controls from May 2004 to July 2007, CT is control, L+ is litter addition, and L- is litter removal.

The estimated annual soil respiration rates were 15.3 t C ha^{-1} yr^{-1}, 19.0 t C ha^{-1} yr^{-1}, and 13.7 t C ha^{-1} yr^{-1} for the CT, L+, and L- treatments, respectively. Thus, the soil carbon lost to the atmosphere in litter addition treatment is at least 4.4 t C ha^{-1} yr^{-1} and may be as high as 6.5 t C ha^{-1} yr^{-1} (23% and 34%, respectively, of expected soil respiration in the L+ plots). Laboratory incubations have demonstrated that repeated additions of fresh organic matter to soil induce greater priming effects than a single addition [37, 38] and that increased decomposition of soil organic matter continued even when the added fresh organic matter had been completely depleted [39]. We therefore expect that chronic increases in litterfall will induce a substantial release of soil carbon in the medium term.

Thus, we show for the first time that increased aboveground litter production in response to global climate change may trigger priming effects and convert considerable amounts of soil carbon to atmospheric carbon dioxide.

4.4 RESULTS

4.4.1 Soil Respiration

Soil respiration showed a seasonal pattern with low rates (c. 130 mg C $m^{-2}h^{-1}$) in the dry season and much higher rates (c. 260 mg C $m^{-2}h^{-1}$) in the wet season. There was no effect of litter manipulation during 2003, the 1st year of treatments, but from May 2004 (17 months after litter manipulation commenced), respiration from the mineral soil in the litter removal (L-) plots was on average 20% lower than in the control (CT) plots (Figure 5).

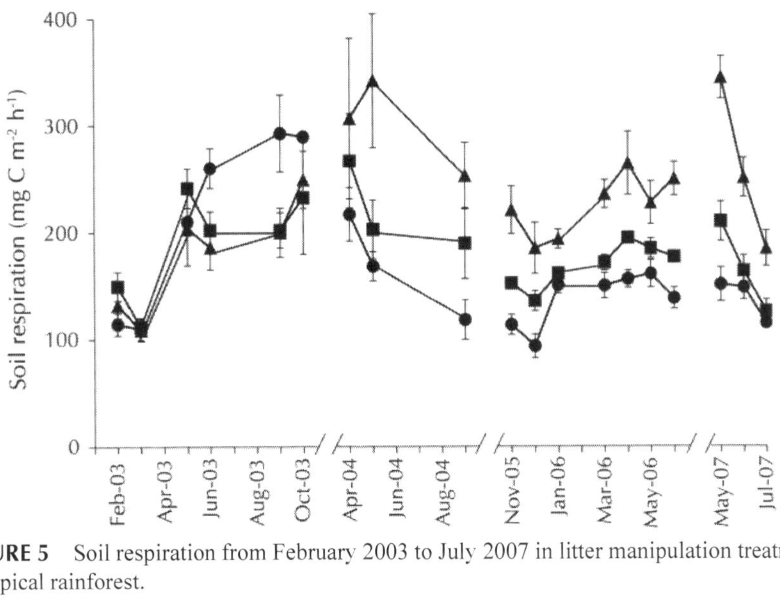

FIGURE 5 Soil respiration from February 2003 to July 2007 in litter manipulation treatments in tropical rainforest.

Soil respiration rates were measured over bare soil in all treatments, squares are controls, triangles are litter addition treatments, and circles are litter removal treatments, error bars show standard errors of means for n = 5.

From the 2nd year of treatments, litter addition increased respiration from the mineral soil more than litter removal decreased it. On average, soil respiration in the litter addition (L+) plots was 43% higher than in the controls and the increase was significant or marginally significant in 11 out of 12 months while respiration in the L- plots was significantly lower than the CT plots in only 2 out of 12 months (Figure 2). The smallest increase in the L+ plots relative to the CT plots was during the dry season (20% in January 2006, Figure 2), while the greatest increases were observed during the dry-to-rainy season transition (69% in May 2004 and 64% in May 2007, Figure 2). The strong increase in soil respiration in the L+ plots was sustained until the end of the study in July 2007 (Figure 5).

The differences are calculated as a percentage of the average respiration measured in the control plots for each month, gray bars are litter addition plots and white bars are litter removal plots, a star above a bar denotes a significant treatment effect ($P< 0.05$) compared to the controls, a circle above a bar denotes a marginally significant treatment effect ($P< 0.065$) compared to the controls.

4.4.2 Soil Temperature and Soil Water Content

Soil temperature (0–100 mm depth) varied very little throughout the year, litter removal decreased soil temperature by c. 0.5°C relative to the CT and L+ plots during the rainy season only in 2003 and 2004 ($P = 0.002$); soil temperature did not differ between the L+ and CT plots except in June and July 2007, when it was 0.3°C ($P = 0.019$) and 0.4°C higher ($P = 0.002$), respectively, in the L+ plots. Soil water content from 0–60 mm depth was not affected by litter manipulation in any season or year.

4.4.3 Fine Root and Microbial Biomass

Fine root biomass in the mineral soil (0–100 mm depth) was 37% lower in the L+ plots than in the CT plots in June and July 2004, after 19 months of litter addition and removal treatments ($P< 0.01$, Figure 1) and 28% lower in August 2006, after 41 months ($P = 0.05$, Figure 1). There was no significant difference in fine root biomass between CT and L- plots in either year.

The CT is control, L+ is litter addition, and L- is litter removal; error bars show standard errors of means for n = 5, different letters above bars indicate a significant difference between treatments at $P< 0.05$. Data for 2004 has been published in a different form [25].

The total microbial C and N (0–100 mm depth) had decreased by 23% in the L- plots relative to the control treatment in August 2004 ($P = 0.011$ and $P = 0.003$ for C and N, respectively, Figure 3(a)) and microbial N in the L- plots was 18% lower than in the CT plots in June 2006 ($P = 0.006$, Figure 3(b)).

Data are given as microbial carbon and nitrogen in (a) August 2004, and (b) June 2006, CT is control, L+ is litter addition, L- is litter removal, error bars show standard errors of means for n = 5, different letters above bars indicate a significant difference between treatments at $P< 0.05$.

Litter addition had no significant effect on microbial biomass C and N in either year.

KEYWORDS

- **Carbon dynamics**
- **Microbial biomass**
- **Soil respiration**
- **Soil water content**
- **Total organic carbon**
- **Tropical litterfall**

AUTHORS' CONTRIBUTIONS

Conceived and designed the experiments: Emma J. Sayer and Edmund V. J. Tanner performed the experiments: Emma J. Sayer analyzed the data: Emma J. Sayer contributed reagents/materials/analysis tools: Jennifer S. Powers wrote the paper: Emma J. Sayer, Jennifer S. Powers, and Edmund V. J. Tanner.

ACKNOWLEDGMENT

We thank Dr. S. Joseph Wright for the use of the Li 6400, Ruscena Wiederholt, Fabienne Zeugin, Jesus Antonio Valdez, Geraldino Perez, Francisco Valdez, Didimo Ureña, and Arturo Worrell Jr. for help in the field, and Milton Garcia for technical advice. All fieldwork was carried out at the Smithsonian Tropical Research Institute, to whom we are very grateful. We also thank B. L. Turner and W. H. Schlesinger for comments on this manuscript.

REFERENCES

1. DeLucia, E. H., Hamilton, J. G., Naidu, S. L., Thomas, R. B., Andrews, J. A., et al. Net primary production of a forest ecosystem with experimental CO_2 enrichment. *Science*, **284**, 1177–1179 (1999).
2. Allen, A. S., Andrews, J. A., Finzi, A. C., Matamala, R., Richter, D. D., and Schlesinger, W. H. Effects of free-air CO_2 enrichment (FACE) on belowground processes in a Pinus taeda forest. *Ecol. Appl.*, **10**, 437–448 (2000).
3. Schlesinger, W. H. and Lichter, J. Limited carbon storage in soil and litter of experimental forest plots under increased atmospheric CO_2. *Nature*, **411**, 466–469 (2001).
4. Finzi, A. C., Allen, A. S., DeLucia, E. H., Ellsworth, D. S., and Schlesinger, W. H. Forest litter production, chemistry, and decomposition following two years of free-air CO_2 enrichment. *Ecology*, **82**, 470–484 (2001).
5. Zak, D. R., Holmes, W. E., Finzi, A. C., Norby, R. J., and Schlesinger, W. H. Soil nitrogen cycling under elevated CO_2: a synthesis of forest FACE experiments. *Ecol. Appl.*, **13**, 1508–1514 (2001).
6. Zhang, X., Zwiers, F. W., Hegerl, G. C., Lambert, H., Gillett, N. P., et al. Detection of human influence on twentieth-century precipitation trends. *Nature*, **448**, 462–465 (2007).
7. Raich, J. W., Russell, A. E., Kitayama, K., Parton, W. J., and Vitousek, P. M. Temperature influences carbon accumulation in moist tropical forests. *Ecology*, **87**, 76–87 (2006).
8. Bardgett, R. D. and Wardle, D. A. Herbivore mediated linkages between aboveground and belowground communities. *Ecology*, **84**, 2258–2268 (2003).

9. Sayer, E. J. Using experimental manipulation to assess the roles of leaf litter in the functioning of forest ecosystems. *Biol. Rev.*, **81**, 1–31(2006).
10. Dixon, R. K., Brown, S., Houghton, R. A., Solomon, A. M., Trexler, M. C., and Wisniewski, J. Carbon pools and flux of global forest ecosystems. *Science*, **263**, 185–190 (1994).
11. Bernoux, M., Carvalho, M. S., Volkoff, B., and Cerri, C. C. Brazil's soil carbon stocks. *Soil Sci. Soc. Am. J.*, **66**, 888–896 (2002).
12. Clark, D. A. Sources or sinks? The responses of tropical forests to current and future climate and atmospheric composition. Philos. *T. Roy. Soc. B*, **359**, 477–491 (2004).
13. Jobbagy, E. G. and Jackson, R. B. The vertical distribution of soil organic carbon and its relation to climate and vegetation. *Ecol. Appl.*, **10**, 423–436 (2000).
14. Raich, J. W., Potter, C. S., and Bhagawati, D. Interannual variability in global soil respiration, 1980–94. *Glob. Change Biol.*, **8**, 800–812 (2002).
15. Palmroth, S., Oren, R., McCarthy, H. R., Johnsen, K. H., Finzi, A. C., et al. Aboveground sink strength in forests controls the allocation of carbon below ground and its [CO_2]-induced enhancement. *Proc. Natl. Acad. Sci. USA*, **103**, 19362–19367 (2006).
16. Kirschbaum, M. U. F. Will changes in soil organic carbon act as a positive or negative feedback on global warming? *Biogeochemistry*, **48**, 21–51 (2000).
17. Davidson, E. A. and Janssens, I. A. Temperature sensitivity of soil carbon decomposition and feedbacks to climate change. *Nature*, **440**, 165–173 (2006).
18. Neff, J. C., Townsend, A. R., Gleixner, G., Lehman, S. J., Turnbull J., and Bowman, W. D. Variable effects of nitrogen additions on the stability and turnover of soil carbon. *Nature*, **419**, 915–917 (2002).
19. Reich, P. B., Hungate, B. A., and Luo, Y. Carbon-nitrogen interactions in terrestrial ecosystems in response to rising atmospheric carbon dioxide. *Annu. Rev. Ecol. Evol. Syst.*, **37**, 611–636 (2006).
20. Cleveland, C. C. and Townsend, A. R. Nutrient additions to a tropical rain forest drive substantial soil carbon dioxide losses the atmosphere. *Proc. Natl. Acad. Sci. USA*, **103**, 10316–10321 (2006).
21. Raich, J. W. and Nadelhoffer, K. J. Belowground carbon allocation in forest ecosystems: global trends. *Ecology*, **70**, 1346–1354 (1989).
22. Davidson, E. A., Savage, K., Bolstad, P., Clark, D. A., Curtis, P. S., et al. Belowground carbon allocation in forests estimated from litterfall and IRGA-based soil respiration measurements. *Agr. Forest Meteorol.*, **113**, 39–51 (2002).
23. Cornejo, F. H., Varela, A., and Wright, S. J. Tropical forest litter decomposition under seasonal drought: nutrient release, fungi and bacteria. *Oikos*, **70**, 183–190 (1994).
24. Vasconcelos, S. S., Zarin, D. J., Capanu, M., Littell, R., Davidson, E. A. et al. Moisture and substrate availability constrain soil trace gas fluxes in an eastern Amazonian regrowth forest. *Global Biogeochem. Cy.*, **18**, GB2009 (2004).
25. Li, Y., Xu, M., Sun, O. J., and Cui, W. Effects of root and litter exclusion on soil CO_2 efflux and microbial biomass in wet tropical forests. *Soil Biol. Biochem.*, **36**, 2111–2114 (2004).
26. Priess, J. A. and Folster, H. Microbial properties and soil respiration in submontane forests of Venezuelan Guyana characteristics and response to fertilizer treatments. *Soil Biol. Biochem.*, **33**, 503–509 (2001).
27. Sayer, E. J., Tanner, E. V. J., and Lacey, A. L. Effects of litter manipulation on early-stage decomposition and meso-arthropod abundance in a tropical moist forest. *Forest Ecol. Manage.*, **229**, 285–293 (2006).
28. Sayer, E. J., Tanner, E. V. J., and Cheesman, A. W. Increased litterfall changes fine root distribution in a moist tropical forest. *Plant Soil*, **281**, 5–13 (2006).
29. Pregitzer, K. S., Laskowski, M. J., and Burton, A. J. Variation in sugar maple root respiration with root diameter and soil depth. *Tree Physiol.*, **18**, 665–670 (1998).
30. Desrochers, A., Landhausser, S. M., and Lieffers, V. J. Coarse and fine root respiration in aspen (*Populus tremuloides*). *Tree Physiol.*, **22**, 725–732 (2002).
31. Singh, J. S. and Gupta, S. R. Plant decomposition and soil respiration in terrestrial ecosystems. *Bot. Rev.*, **43**, 449–528 (1977).

32. Anderson, J. M., Proctor, J., and Vallack, H. W. Ecological studies in four contrasting lowland rain forests in Gunung-Mulu National Park, Sarawak. 3. Decomposition processes and nutrient losses from leaf litter. *J. Ecol.*, **71**, 503–527 (1983).
33. Rout, S. K. and Gupta, S. R. Soil respiration in relation to abiotic factors, forest floor litter, root biomass and litter quality in forest ecosystems of Siwaliks in northern India. *Acta Oecol.*, **10**, 229–244 (1989).
34. Metcalfe, D. B., Meir, P., Aragão, L. E. O. C., Mahli, Y., Da Costa, A. C. L., et al. Factors controlling spatio-temporal variation in carbon dioxide efflux from surface litter, roots, and soil organic matter at four rain forest sites in eastern Amazon. *J. Geophys. Res.*, **112**, G04001 (2007).
35. Bingeman, C. W., Varner, J. E., and Martin, W. P. The effect of the addition of organic materials on the decomposition of an organic soil. *Soil Sci. Soc. Am. Proc.*, **29**, 692–696 (1953).
36. Kuzyakov, Y., Friedel, J. K., and Stahr, K. Review of mechanisms and quantification of priming effects. *Soil Biol. Biochem.*, **32**, 1485–1498 (2000).
37. DeNobili, M., Contin, M., Mondini, C., and Brookes, P. C. Soil microbial biomass is triggered into activity by trace amounts of substrate. *Soil Biol. Biochem.*, **33**, 1163–1170 (2001).
38. Hamer, U. and Marschner, B. Priming effects in soils after combined and repeated substrate additions. *Geoderma*, **128**, 38–51 (2005).
39. Fontaine, S., Bardoux, G., Abbadie, L., and Mariotti, A. Carbon input to soil may decrease soil carbon content. *Ecol. Lett.*, **7**, 314–320 (2004).
40. Cavalier, J. Fine-root biomass and soil properties in a semi-deciduous and a lower montane rain forest in Panama. *Plant Soil*, **142**, 187–201 (1992).
41. Powers, J. S., Treseder, K. K., and Lerdau, M. T. Fine roots, arbuscular mycorrhizal hyphae and soil nutrients in four neotropical rain forests: patterns across large geographic distances. *New Phytol.*, **165**, 913–921 (2005).
42. Leigh, E. G. *Tropical Forest Ecology*. Oxford University Press, Oxford (1999).
43. Leigh, E. G. and Wright, S. J. *Barro Colorado Island and tropical biology*. In: A. H Gentry (Ed.). Four Neotropical Rainforests. Yale University Press, New Haven, pp. 28–47 (1990).
44. Brookes, P. C., Landman, A., Pruden, G., and Jenkinson, D. S. Chloroform fumigation and the release of soil nitrogen a rapid direct extraction method to measure microbial biomass nitrogen in the soil. *Soil Biol. Biochem.*, **17**, 837–842 (1985).
45. Beck, T., Joergensen, R. G., Kandeler, E., Makeschin, F., Nuss, E., et al. An inter-laboratory comparison of ten different ways of measuring soil microbial biomass C. *Soil Biol. Biochem.*, **29**, 1023–1032 (1997).

5 Altitudinal Variations in Soil Carbon

Mehraj A. Sheikh, Munesh Kumar, and Rainer W. Bussmann

CONTENTS

5.1 INTRODUCTION

The Himalayan zones, with dense forest vegetation, cover a fifth part of India and store a third part of the country reserves of soil organic carbon (SOC). However, the details of altitudinal distribution of these carbon stocks which are vulnerable to forest management and climate change impacts are not well known.

This study reports the results of measuring the stocks of SOC along altitudinal gradients. The study was carried out in the coniferous subtropical and broadleaf temperate forests of Garhwal Himalaya. The stocks of SOC were found to be decreasing with altitude: from 185.6 to 160.8 t C ha^{-1} and from 141.6 to 124.8 t C ha^{-1} in temperature (*Quercus leucotrichophora*) and subtropical (*Pinus roxburghii*) forests, respectively.

The results of this study lead to conclusion that the ability of soil to stabilize soil organic matter (SOM) depends negatively on altitude and call for comprehensive theoretical explanation.

Soils are the largest carbon reservoirs of the terrestrial carbon cycle. About three times more carbon is contained in soils than in the world's vegetation and soils hold double the amount of carbon that is present in the atmosphere. Worldwide the first 30 cm of soil holds 1500 pg carbon [1], for India the figure is 9 pg [2]. Soils play a key role in the global carbon budget and greenhouse effect [3]. Soils contain 3.5% of the earth's carbon reserves compared with 1.7% in the atmosphere, 8.9% in fossil fuels, 1.0% in biota, and 84.9% in the oceans [4]. The amount of CO_2 in the atmosphere steadily increases as a consequence of anthropogenic emissions, but there is a large interannual variability caused by terrestrial biosphere [5].

The first estimate of the organic carbon (OC) stock in Indian soils was 24.3 pg (1 pg = 1015 g) based on 48 soil samples [6]. Forest soils are one of the major carbon sinks on earth, because of their higher organic matter content [7]. Soils can act as sinks or as a source for carbon in the atmosphere depending on the changes happening to SOM. Equilibrium between the rate of decomposition and rate of supply of organic matter is disturbed when forests are cleared and land use is changed [8, 9]. The SOM can also increase or decrease depending on numerous factors, including climate, vegetation type, nutrient availability, disturbance, land use, and management practice [10, 11]. Physical soil properties, such as soil structure, particle size, and composition have profound impact on soil carbon (C). Soil particle size has an influence on the rate of decomposition of SOC [12]. The release of nutrients from litter decomposition is a fundamental process in the internal biogeochemical cycle of an ecosystem and decomposers recycle a large amount of carbon that was bounded in the plant or tree to the atmosphere [13].

About 40% of the total SOC stock of the global soils resides in forest ecosystem [14]. The Himalayan zones, with dense forest vegetation, cover nearly 19% of India and contain 33% of SOC reserves of the country [15]. These forests are recognized for their unique conservation value and richness of economically important biodiversity. Managing these forests may be useful technique to increase soil carbon status because the presence of trees affects carbon dynamics directly or indirectly. Trees improve soil productivity through ecological and physico-chemical changes that depend upon the quantity and quality of litter reaching soil surface and rate of litter decomposition and nutrient release [16].

The current global stock of SOC is estimated to be 1,500–1,550 pg [1, 17, 18]. This constituent of the terrestrial carbon stock is twice that in the earth's atmosphere (720 pg), and more than triple the stock of OC in terrestrial vegetation (560 pg) [19, 20]. To sustain the quality and productivity of soils, knowledge of SOC in terms of its amount and quality is essential. The first comprehensive study of OC status in Indian soils was conducted [21] by collected 500 soil samples from different cultivated fields and forests with variable rainfall and temperature patterns. However, the study did not make any estimate of the total carbon reserves in the soils. The first attempt in estimating OC stock [22] was also made based on a hypothesis of enhancement of OC level on certain unproductive soils. In last decade, the greenhouse effect has been of great concern and has led to several studies on the quality, kind, distribution, and behavior of SOC [1, 23, 24]. Global warming and its effect on soils in terms of SOC management have led to several quantitative estimates for global C content in the soils [1-26]. Although, so far the SOC stock studies in Indian Himalayan forests in relation to altitudinal gradient are not available. Therefore, the aim of the present study is made to estimate SOC stocks of two dominant forests of subtropical (*Pinus roxburghii*) and temperate (*Quercus leucotrichophora*) along altitudinal gradient in Garhwal Himalaya.

5.2 METHODS

The study area is situated in Tehri Garhwal, one of the western most districts of the Uttarakhand State and located on the outer ranges of the Mid-Himalayas which comprise low line peaks rising directly from the plains of the Northern India. The study site lies between 30° 18' 15.5" and 30° 20' 40" N latitude and 78° 40' 36.1" to 78° 37' 40.4" E longitude. Three sites were selected within *Pinus roxburghii* forest at an altitude of 700 m (site-I), 900 m (site-II), 1100 m asl (site-III), and three sites in *Quercus leucotrichophora* forest at altitudes of 1700 m (site-I), 1900 m (site-II), and 2100 m (site-III).

The quality of OC data of the soils depends on sampling methods, the kind of vegetation, and the method of soil analysis performed in the laboratory. The sampling was done by nested plot design method. In each site, a plot of 100×20 m^2 size was laid, and six sampling points were selected in each plot by the standard method [57]. Three samples were collected at each sampling point at three depths (0–20, 20–40, and 40–60 cm). A total of 108 soil samples (18 from each site) were collected by digging soil pits ($6 \times 3 \times 6$ cm^3). The soil samples were air dried and sieved (< 2 mm) before analysis. The SOC for various depths was determined by partial oxidation method [58]. Soil samples from each depth were analyzed, however to express the total SOC stock data in 0–20, 20–40, and 40–60 cm, the weighted mean average were considered. The total SOC stock was estimated by multiplying the values of SOC g kg^{-1} by a factor of 8 million, based in the assumption that a layer of soil 60 cm deep covering an area of 1 ha weighs 8 million kg [7].

5.3 DISCUSSION

The SOC decreased with increasing soil depths in all the sites except site-II of the *Pinus roxburghii* forest, where OC was highest in the top layer (0–20 cm) and lowest in middle depth (20–40 cm). In the *Quercus leucotrichophora* forest for all sites (site-I, site-II, and site-III) the level of SOC was higher in the upper layer, dropping with an increase in depth. The similar trend (higher in top layer and decreased with increasing depths) of SOC is also reported in the *Pinus roxburghii*. The higher OC content in the top layer may be due to rapid decomposition of forest litter in a favorable environment. The SOC represents [27] a significant pool of carbon within the biosphere. Climate shifts in temperature and precipitation have a major influence on the decompositions and amount of SOC stored within on ecosystem and that released into the atmosphere. The rate of cycling of carbon at different depths and in different pools across different vegetal cover is still not clear. There is not, as yet, enough information to predict how the size and residence time of different fractions of SOC varies [28]. The higher concentration of SOC in top layer has also been reported by various authors [28, 29]. The steep fall in the SOC content as depth increases is an indication of higher biological activity associated with top layers. Our results are in accordance with earlier studies [28, 30].

The maximum carbon stock was present in *Quercus leucotrichophora* forest soils. The higher percent of SOC in *Quercus leucotrichophora* forest may be due to dense canopy and higher input of litter which results in maximum storage of carbon stock.

In *Quercus leucotrichophora* forest sites dense vegetation led to higher accumulation of SOC as compared to coniferous sites. The higher accumulation of SOC found in maquis vegetation, as opposed to coniferous forest has been reported [31]. In *Pinus roxburghii* forest, the lower amount of OC might be due to wider spacing between trees, resulting in lower litter input and less accumulation in turn yielding less storage of carbon stock in these forest soils. The study of [32] indicated a positive influence of residue application on soil carbon content. The added litter [33] and the proliferated root system [34] of the growing plants probably influenced the carbon storage in the soil, suggesting a positive correlation of SOC with the quantity of litter fall [35]. The study [36] suggested that coarse and fine woody debris are substantial forest ecosystem carbon stock. The production and decay rate of forest woody detritus depends partially on climatic conditions. The results of this study indicated that highest carbon stock founding region with cool summer, while lower carbon in arid desert/steppes or temperate humid regions.

In *Quercus leucotrichophora* forest soils (Table 1), the maximum and minimum values of carbon stock was 185.6 t C ha^{-1} (site-I) and 160.8 t C ha^{-1} (site-III), respectively. The trend was similar for the *Pinus roxburghii* forest soils (Table 2), where the highest and lowest values of carbon stock was 141.6 t C ha^{-1} (site-I) and 124.8 t C ha^{-1} (site-III). A study of [37] recorded the following levels of OC stored in some Indian soils: 41.2 t C ha^{-1}, 120.4 t C ha^{-1}, 13.2 t C ha^{-1}, and 18 t C ha^{-1} in the red soil, laterite soil, saline soil, and black soil, respectively. All these measurements were lower than in the present study. Another study showed [3] the national average content of SOC was 182.94 t C ha^{-1}. The total amount of SOC stored in *Quercus leucotrichophora* forest soils is almost similar to the national average and expresses the excellent ability of these forests to stock and sequester organic carbon. However, the total amount of OC stored in *Pinus roxburghii* forest soils was lower than the national average.

TABLE 1 The SOC stock (up to 60 cm depth) in *Quercus leucotrichophora* forest.

Site	Altitudinal range	SOC g kg^{-1}	Carbon stock (t C ha^{-1})
Site-I	1,600–1,800 m	23.2 ± 1.2	185.6
Site-II	1,800–2,000 m	22.6 ± 0.7	180.8
Site-III	2,000–2,200 m	20.1 ± 3.2	160.8

TABLE 2 The SOC stock (up to 60 cm depth) in *Pinus roxburghii* Forest.

Site	Altitudinal range	SOC (%)	Carbon stock (t C ha^{-1})
Site-I	600–800 m	17.7 ± 0.24	141.6
Site-II	800–1,000 m	15.8 ± 0.42	126.4
Site-III	1,000–1,200 m	15.6 ± 0.31	124.8

A study carried of grassland in two different sites that is Mehrstedt and Kaltenborn, where SOC stocks at the clay rich Mehrstedt site were almost twice as high as at the sandy Kaltenborn site [38]. The clay soil texture was contained on average 123 t C ha^{-1} for 0–60 cm depth. A compilation of 121 soil profiles of temperate grasslands, mainly from North America from several databases resulted in a mean carbon stock of 91 t C ha^{-1} for 0–60 cm depth [39]. However, the range of carbon stocks in temperate grasslands may be between 30 and 80 t C ha^{-1} [40]. The SOM is a major component of global carbon cycle [41], increases with precipitation and decreases with temperature [42-44]. The SOM content were also reported in the top 0–50 cm soil layer is positively correlated with the precipitation/temperature ratio in the pampa and chaco soils in Argentina [45].

While comparing the SOC stock values of different sites with each other in both forests, the carbon stock tended to decrease with increasing altitudes. A soil carbon study in Kathmandu valley of Nepal in *Pinus roxburghii* forest along altitudinal gradient at an elevation ranging between 1,200 and 2,200 reported that the higher altitude soil was found to be much more depleted of C than the lower altitude soil [46]. The decreasing trend of C might be attributed to the lower mineralization rate and net nitrification rate at the higher altitude. A study carried out [47] in Himalayan forests indicates a characteristic decline in total tree density and basal area was apparent with increasing altitude. In the present study, a characteristic decline in vegetation was observed across altitudinal strata and among sites. The decrease in species richness in high elevation strata could be due to eco-physiological constraints, low temperature, and productivity [48]. Altitude had a significant effect on species richness which declines with even a 100 m increase in altitude. Species composition too is significantly affected by altitude [49]. Altitude is often employed to study the effects of climatic variables on SOM dynamics [44, 50]. Temperature decreased and precipitation increased with increasing altitude. The changes in climate along altitudinal gradients influence the composition and productivity of vegetation and consequently, affect the quantity and turnover of SOM [50, 51]. Altitude also influences SOM by controlling soil water balance, soil erosion, and geologic deposition processes [52]. The advantages of altitudinal gradients in forest soil for testing the effects of environmental variables on SOM dynamics is emphasized [50]. The relationship between SOM and altitude has also been investigated and positive correlations were reported [53, 54]. A study of wetland, the balance between carbon input (organic matter production) and output (decomposition, methanogenisis, etc.) and the resulting storage of carbon depend on topography and the geological position of wetland, the hydrological regime, the type of plant present, the temperature, moisture of the soil, pH, and the morphology [55]. There is a strong relation between climate and soil carbon pools where OC content decreases with increasing temperatures, because decomposition rates doubles with every 10°C increase in temperature [41].

The characteristic decline in vegetation with increasing altitude results in less accumulation of litter and low input of OC in soils. Similar findings were also reported [13], the number of trees per hectare decreases with increasing elevation, the comments related to kg ha^{-1} unquestionable give consequences implying that all weight parameters decreases at the altitude increases. A study carried out in the Western Ghats of Southern India also shows the decline of SOC from 110.2 t C ha^{-1} at an elevation >1400 m to 82.6 t C ha^{-1} at an elevation >1800 m [56]. The increasing tendency of carbon density with decreasing altitude may be better stabilization of SOC at lower altitudes. It is a proven fact that forest ecosystems are the best way to sequester carbon however, considering the huge human population in developing country like India, much of the land cannot be spared for increase in forest cover. In such circumstance the management of vast areas of Himalayan forests at lower elevations can be regarded as major sinks of mitigating atmospheric carbon dioxide. Forests at higher altitudes can be seen as potential carbon sinks.

5.4 RESULTS

Depth-wise SOC results are mentioned in Tables 3 and 4. A decreasing trend in SOC was observed with increased soil depths in all the sites except site-II of the *Pinus roxburghii* forest, where OC was highest in the top layer (0–20 cm) and lowest in middle depth (20–40 cm). The carbon level increased again below the middle depth. In site-I of the *Quercus leucotrichophora* forest, the level of SOC ranged from 24.3 ± 1.9 to 21.9 ± 3.1 g kg^{-1} and was higher in the upper layer, dropping with an increase in depth. The trend was same for site-II and site-III where the SOC values also decreased with increasing depths and ranged from 23.4 ± 0.8 to 21.9 ± 1.2 g kg^{-1} and 22.5 ± 2.6 to 16.5 ± 2.1 g kg^{-1}, respectively. The range of SOC in *Pinus roxburghii* forest was 18.0 ± 6.5 to 12.1 ± 0.9 g kg^{-1}, 19.6 ± 0.9 to 11.2 ± 0.3 g kg^{-1}, and 19.6 ± 0.5 to 15.0 ± 0.2 g kg^{-1} for site-I, site-II, and site-III, respectively, again the levels were higher in the top layer and decreased with depth.

TABLE 3 The SOC (± SD) values at different depths of *Quercus leucotrichophora* forest soils.

Site	Soil depth (cm)	SOC g kg^{-1}
Site-I	0–20	24.3 ± 1.9
	20–40	23.4 ± 3.4
	40–60	21.9 ± 3.1
Site-II	0–20	23.4 ± 0.8
	20–40	22.5 ± 3.3
	40–60	21.9 ± 1.2
Site-III	0–20	22.5 ± 2.6
	20–40	21.5 ± 6.8
	40–60	16.5 ± 2.1

TABLE 4 The SOC (± SD) values at different depths in *Pinus roxburghii* forest soils.

Site	Soil depth (cm)	SOC g kg^{-1}
Site-I	0–20	18.0 ± 6.5
	20–40	16.8 ± 4.1
	40–60	12.1 ± 0.9
Site-II	0–20	19.6 ± 0.9
	20–40	11.2 ± 0.3
	40–60	16.8 ± 5.3
Site-III	0–20	18.0 ± 6.5
	20–40	18.7 ± 8.4
	40–60	15.0 ± 0.2

The maximum carbon stock was present in *Quercus leucotrichophora* forest soils. The higher percent of SOC in *Quercus leucotrichophora* forest may be due to dense canopy and higher input of litter which results in maximum storage of carbon stock. In *Quercus leucotrichophora* forest sites dense vegetation led to higher accumulation of SOC as compared to coniferous sites. In *Pinus roxburghii* forest, the lower amount of OC might be due to wider spacing between trees, resulting in lower litter input, and less accumulation in turn yielding less storage of carbon stock in these forest soils.

In *Quercus leucotrichophora* forest soils (Table 1), the maximum carbon stock was present in site-I (185.6 t C ha^{-1}) and minimum in site-III (160.8 t C ha^{-1}). The trend was the same for the *Pinus roxburghii* forest soils (Table 2), where the highest carbon stock was present in site-I (141.6 t C ha^{-1}) followed by site-II (126.4 t C ha^{-1}), and site-III (124.8 t C ha^{-1}). While comparing the SOC stock values of different sites with each other in both forests, the carbon stock tended to decrease with increasing altitudes. In the present study, a characteristic decline in vegetation was observed across altitudinal strata and among sites. Altitude had a significant effect on species richness, which declines with even a 100 m increase in altitude. The characteristic decline in vegetation with increasing altitude results in less accumulation of litter and low input of OC in soils.

5.5 CONCLUSION

A comparison of the SOC stock values of different sites in both forests show that the carbon stock tons per hectare decrease with increasing altitudes. The tendency of carbon density to increase as altitude decreases may be due to better stabilization of SOC at lower altitudes. Considering the huge human population in developing country like

India, much of the land cannot be spared for increase in forest cover. In such circumstance the management of vast areas of Himalayan forests at lower elevations can be regarded as major sinks of mitigating atmospheric carbon dioxide.

KEYWORDS

- **Carbon density**
- **Ecosystem**
- **Global warming**
- **Organic matter content**
- **Soil organic carbon**
- **Soil organic matter**

AUTHORS' CONTRIBUTIONS

Mehraj A. Sheikh and Munesh Kumar share the contributions to fieldwork, data analysis, and compilation of this manuscript. Rainer W. Bussmann shares his valuable contribution from initial manuscript drafting to final submission. All authors read and approved the final manuscript.

ACKNOWLEDGMENT

The authors are thankful to the reviewers and the Handling Editor for their constructive comments and suggestions.

REFERENCES

1. Batjes, N. H. Total C and N in soils of the world. *Eur. J. Soil Sci.*, **47**, 151–163 (1996).
2. Bhattacharya, T., Pal, D. K., Mondal, C., and Velayutham, M. Organic carbon stock in Indian soils and their geographical distribution. *Current Science*, **79**, 655–660 (2000).
3. Jha, M. N., Gupta, M. K., Saxena, A., and Kumar, R. Soil organic carbon store in different forests in India. *Indian Forester*, **129**, 714–724 (2003).
4. Lal, R., Kimble, J. M., and Levines, E. Whiteman C. *World soil and greenhouse effect*. SSSA Special Publication Number Madison, Wisconsin **57**, 51–65 (1995).
5. Erbrecht, T. and Lucht, W. Impact of large scale climatic disturbances on the terrestrial carbon cycle. *Carbon Balance and Management*, **1**, 7 (2006).
6. Gupta, R. K. and Rao, D. L. N. Potential of wastelands for sequestering carbon by reforestation. *Current Science*, **66**, 378–380 (1984).
7. Dey, S. K. A preliminary estimation of carbon stock sequestrated through rubber (*Hevea brasiliensis*) plantation in North Eastern regional of India. *Indian Forester*, **131**, 1429–1435 (2005).
8. Lal, R. Soil carbon sequestration to mitigate climate change. *Geoderma*, **123**, 1–22 (2004).
9. Buringh, P. Organic carbon in soils of world. In: G. M. Woodwell. (Ed.). *The role of terrestrial vegetation in global carbon cycle Measurement by remote sensing*. John Wiley, New York, pp. 91–109 (1984).
10. Six, J. and Jastrow, J. D. Organic matter turnover. *Encycl. Soil Sci.*, pp. 936–942 (2002).
11. Baker, D. F. Reassessing carbon sinks. *Science*, **316**, 1708–1709 (2007).
12. Jobbagy, E. G. and Jackson, R. B. The vertical distribution of soil organic and its relation to climate and vegetation. *Ecology Application*, **10**, 423–426 (2000).

13. Sevgi, O. and Tecimen, H. B. Changes in Austrian Pine forest floor properties in relation with altitude in mountainous areas. *Journal of Forest Science*, **54**, 306–313 (2008).
14. Eswaran, H., Reich, P., Kimble, J. M., Beinroth, F. H., Padmanabhan, E., and Moncharoen, P. In: Global Climate Change and Pedogenic Carbonates. R. Lal., et al. (Eds.). Lewis Publishers, Miami, Florida, pp. 15–25 (1999).
15. Bhattacharyya, T., Pal, D. K., Chandran, P., Ray, S. K., Mandal, C., and Telpande, B. Soil carbon storage capacity as a tool to prioritize areas for carbon sequestration. *Current Science*, **95**, 482–494 (2008).
16. Meentemeyer, V. and Berg, B. Regional variation in rate of mass loss of *Pinus sylvestris* needle litter in Swedish pine forest as influenced by climate and litter quality. *Scand J. For. Res.*, **1**, 167–180 (1986).
17. Lal, R. Soil carbon sequestration impacts on global change and food security. *Science*, **304**, 1623–1627 (2004).
18. Post, W. M., Izaurralde, R. C., Mann, L. K., and Bliss, N. Monitoring and verifying changes of organic carbon in soil. *Climate Change*, **51**, 73–99 (2001).
19. Baes, C. F. Jr., Goeller, H. E., Olson, J. S., and Rotty, R. M. Carbon dioxide and climate the uncontrolled experiment. *Am. Sci.*, **65**, 310–320 (1977).
20. Bolin, B. The carbon cycle. *Sci. Am.*, **233**, 124–132 (1970).
21. Jenny, H. and Raychaudhuri, S. P. *Effect of Climate and Cultivation on Nitrogen and Organic Matter Reserves in Indian Soils*. ICAR, New Delhi, India, p. 126 (1960).
22. Gupta, R. K. and Rao, D. L. Potential of wastelands for sequestering carbon by reforestation. *Current Science*, **66**, 378–380 (1994).
23. Eswaran, H., Van, D. B., and Reich, P. Organic carbon in soils of the world. *Soil Sci. Soc. Am. J.*, **57**, 192–194 (1993).
24. Velayutham, M., Pal, D. K., and Bhattacharyya, T. In *Global Climate Change and Tropical Ecosystems*. R. Lal, J. M. Kimble, and B. A. Stewart. (Eds.). Lewis, Boca Raton, Florida, pp. 71–96 (2000).
25. Buringh, P. In *The Role of Terrestrial Vegetation in the Global Carbon Cycle Measurements by Remote Sensing*. G. M. Woodwell (Ed.). John Wiley, New York, pp. 91–109 (1984).
26. Kimble, J., Cook, T., and Eswaran, H. Proc Symp Characterization and Role of Organic Matter in Different Soils, *14th Soil* Sci, Kyoto, Japan, Wageningen, Netherlands, pp. 250–258 (August, 12–18 1990).
27. Grace, P., Post, W., and Hennessy, K. The potential impact of climate change on Australia's soil organic carbon resources. *Carbon Balance and Management*, **1**, 14 (2006).
28. Dinakaran, J. and Krishnayya, N. S. R. Variation in type of vegetal cover and heterogeneity of soil organic carbon in affecting sink capacity of tropical soils. *Current Science*, **94**, 9 (2008).
29. Alamgir, M. and Amin, M. A. Storage of organic carbon in forest undergrowth, litter and soil within geoposition of Chittagong (south) forest division. *Bangladesh. International Journal of Usufruct. Management*, **9**, 75–91 (2008).
30. Wang, S., Huang, M., Shao, X., Mickler, A. R., LI, K., and Ji, J. Vertical distribution of soil organic carbon in China. *Environ. Manage.*, **33**, 200–209 (2004).
31. Markus E., Ladina, A., Aldo, M., Salvatore, R., Markus, N., and Rene, V. Effect of climate and vegetation on soil organic carbon, humus fractions, allophones, imogolite, kaolinite and oxyhydroxides in volcanic soils of Etna (Sicily). *Soil Science*, **172**, 673–691 (2007).
32. Havelin, J. L., Kissel, D. E., Maddux, L. D., Classe, M. M., and Long, J. H. Crop rotation and tillage effects on soil carbon and nitrogen. *Soil Sci. Soc. Am. J.*, **54**, 448–452 (2003).
33. Lal, R. Conservation tillage for sustainable agriculture tropics versus temperate environment. *Advances in Agronomy*, **42**, 186–197 (1989).
34. Blevines, R. L. and Frye, W. W. Conservation tillage an ecological approach in soil management. *Advances in Agronomy*, **51**, 33–78 (1993).
35. Singh, G. Carbon sequestration under an Agri-silvicultural system in the Arid region. *Indian Forester*, **147**, 543–552 (2005).

36. Woodall, C. and Liknes, G. Climatic regions as an indicator of forest coarse and fine woody debris carbon stocks in the United States. *Carbon Balance and Management*, **3**, 5 (2008).
37. Jha, M. N., Gupta, M. K., and Raina, A. K. Carbon sequestration Forest Soil and Land Use Management. *Annals of Forestry*, **9**, 249–256 (2001).
38. Axel, D., Schumacher, J., Scherer-Lorenzen, M., Scholten, T., and Schulze, E. D. Spatial and vertical variation of soil carbon at two grassland sites—Implications for measuring soil carbon stocks. *Geoderma*, **141**, 272–282 (2007).
39. Jobbagy, E. G. and Jackson, R. B. The vertical distribution of soil organic carbon and its relation to climate and vegetation. *Ecological Applications*, **10**, 423–436 (2000).
40. Conant, R. T. and Paustian, K. Potential soil carbon sequestration in overgrazed grassland ecosystems. *Global Biogeochemical Cycles*, **16**, 1143 (2002).
41. Schlesinger, W. H. *An Analysis of Global Change*. Academic Press, San Diego, Califorina (1997).
42. Jenny, H. *The Soil Resource Origin and Behavior, Ecological Studies. Vol. 37*. Springer-Verlag, New York (1980).
43. Percival, H. J., Parfitt, R. L., and Scott, N. A. Factors controlling soil carbon levels in New Zealand grasslands: is clay content important. *Soil Science Society of America Journal*, **64**, 1623–1630 (2000).
44. Lemenih, M. and Itanna, F. Soil carbon stocks and turnovers in various vegetation type and arable lands along an elevation gradient in southern Ethiopia. *Geoderma*, **123**, 177–188 (2004).
45. Alvarez, R. and Lavado, R. S. Climate, organic matter and clay content relationships in the Pampa and Chaco soils, Argentina. *Geoderma*, **83**, 127–141 (1998).
46. Sah, S. P. and Brumme, R. Altitudinal gradients of natural abundance of stable isotopes of nitrogen and carbon in the needles and soil of a pine forest in Nepal. *Journal of Forest science*, **49**, 19–26 (2003).
47. Gairola, S., Rawal, R. S., and Todaria, N. P. Forest vegetation patterns along an altitudinal gradient in Sub-alpine zone of West Himalaya India. *African Journal of Plant Science*, pp. 42–48 (2004).
48. Korner, C. A re-assessment of high elevation of tree line positions and their explanations. *Oecologia*, **115**, 445–459 (1998).
49. Hardy, F. G., and Syaukani, Eggleton P. The effect of altitude and rainfall on the composition of termites (Isotera) of the Leuser ecosysrem (Sumatra, Indonesia). *Journal of Tropical Ecology*, **17**, 379–393 (2001).
50. Garten, C. T., Post, W. M., Hanson, P. J., and Cooper, L. W. Forest soil carbon inventories and dynamics along an elevation gradient in the southern Appalachian Mountains. *Biogeochemistry*, **45**, 115–145 (1999).
51. Quideau, S. A., Chadwick, Q. A., Benesi, A., Graham, R. C., and Anderson, M. A. A direct link between forest vegetation type and soil organic matter composition. *Geoderma*, **104**, 41–60 (2001).
52. Tan, Z. X., Lal, R., Smeck, N. E., and Calhoun, F. G. Relationships between surface soil organic carbon pool and site variables. *Geoderma*, **21**, 185–187 (2004).
53. Sims, Z. R. and Nielsen, G. A. Organic carbon in Montana soils as related to clay content and climate. *Soil Science Society of America Journal*, **50**, 1269–1271 (1986).
54. Tate, K. R. Assessment, based on a climosequence of soil in tussock grasslands, of soil carbon storage and release in response to global warming. *Journal of Soil Science*, **43**, 697–707 (1992).
55. Shalu Adhikari1, Roshan, M., Bajracharaya1, and Bishal, K. S. A Review of Carbon Dynamics and Sequestration. *Wetlands Journal of Wetlands Ecology*, **2**, 41–45 (2009).
56. Krishnan, P., Bourgeon, G., Seen, D. L., Nair, K. M., Prasanna, R., Srinivas, S., Muthusankar, G., Dufy, L., and Ramesh, B. R. Organic carbon stock map for soils of Southern India. A multifactor approach. *Current Science*, **5**, 10 (2007).

57. Hairiah, K., Sitompul, V. S., Noordwijk, M., and Palm, C. *Methodology for sampling carbon stocks above and below ground*. ASB Lecture Notes 4B. International Centre for Research in Agro forestry, Indonesia (December, 2001). Retrieved from http://www.icraf.cgiar.org/sea
58. Walky, A. and Black, I. A. An examination of the Degtiareff method for deteming soil organic matter and proposed modification of the chromic acid titration method. *Soil Science*, **63**, 29–38 (1934).

6 Carbon Pool Dynamics in Russia's Sandy Soils

Olga Kalinina, Sergey V. Goryachkin, Nina A. Karavaeva, Dmitriy I. Lyuri, and Luise Giani

CONTENTS

6.1 INTRODUCTION

Until recently, a lot of arable lands were abandoned in many countries of the world and, especially, in Russia, where about half a million square kilometers of arable lands were abandoned in 1961–2007. The soils at these fallows undergo a process of natural restoration (or self restoration) that changes the balance of soil organic matter (SOM) supply and mineralization.

A soil chronosequence study, covering the ecosystems of 3, 20, 55, 100, and 170 years of self restoration in southern taiga zone, shows that soil organic content of mineral horizons remains relatively stable during the self restoration. This does not imply, however, that SOM pools remain steady. The C/N ratio of active SOM reached steady state after 55 years, and increased doubly (from 12.5–15.6 to 32.2–33.8). As to the C/N ratio of passive SOM, it has been continuously increasing (from 11.8–12.7 to 19.0–22.8) over the 170 years, and did not reach a steady condition.

The results of the study imply that soil recovery at the abandoned arable sandy lands of taiga is incredibly slow process. Not only soil morphological features of a former ploughing remained detectable but also the balance of SOM input and mineralization remained unsteady after 170 years of self restoration.

Predominantly caused by economic crises, the most abandonment was found in Russia, reaching 578000 km^2 in the years 1961–2007 [1–3]. As a consequence, the soils of these sites underwent the process of natural restoration or self restoration.

A recent chronosequential study on the succession of vegetation, profile morphology, and soil properties of post-agrogenic sandy soils under self restoration of the southern taiga zone in the European part of Russia showed that the vegetation developed towards spruce forest and the soils towards natural Podzols with an accumulation of thick raw humus layers [4]. Additional podzolization features were found in respect to morphology and chemical properties like pH, exchangeable cations, and nutrition dynamics. Although these changes happened rather fast, the ploughing features were still evident after 170 years of self restoration, as found in other studies [1, 2, 5, 6].

Since every land use change causes a disturbance of the long-termed adjusted balance of SOM supply and mineralization, self restoration also leads to alterations in the SOM dynamics. In respect of afforestation of former arable sites, the most evident effect on C sequestration was the net sink of atmospheric CO_2 with C accumulation mainly in the growing trees and the forest floor [7-11]. The mineral soil was found to account for less than 1% of radio carbon accretion [10] and might be even less due to varying interrelations between initial soil organic carbon (SOC) pools and SOC dynamics trends in respect to environmental conditions [12, 13]. The results on C sequestration of post-agrogenic sandy soils under self restoration of the southern taiga zone in the European part of Russia also indicate an overall sink [4]. Decreased SOC stores of the 0.2 m mineral soil were overcompensated by increasing SOC stores of the raw humus layer, the latter rating 10% of the whole SOC stores after 55 years of self restoration, and 40% after 100 years and 52% after 170 years, respectively.

Self restoration does not only affect carbon sequestration but also influences the different functional carbon pools and qualities. These changes in SOM are already indicated by increasing C/N ratios during self restoration of post-agrogenic soils [4].

They are expected not to proceed consistently because they preferentially affect the SOM with a short turnover time corresponding to free particulate organic matter (POM) and the organic matter (OM) of the sand and coarse silt fractions [14-16]. Diminishing plant OM transformation induced an accumulation of particulate SOM from poorly decomposed plant residues producing raw humus in Podzols [11, 17]. Because of low bioturbation, this SOM is considered for the SOM dynamics of the mineral soil. Therefore, during self restoration, decreasing particulate SOM from former land use is expected in the mineral soil and perhaps simultaneously, increasing particulate SOM from fine root litter supplied by newly established plants. The changes in SOM dynamics are also expected for the mineral bound fraction. John et al. [16] reported more mineral bound organic carbon (OC) in cultivated soils than in forest topsoil's. As discrete fractions show different times of decomposition in the range of 20–50 years for particulate SOM and about 100 years or more for mineral associated SOM [15, 16, 18], this decrease is assumed to be a long-term one at the site of investigation.

To obtain additional information on C dynamics and alterations of soils under self restoration, this study was carried out with a focus on different functional carbon pools and their qualities. We used the same soil chronosequence, which has been investigated for changes of basic soil properties [4].

6.2 METHODS

6.2 1 Site of Investigation

The study was done at the same sites where Kalinina et al. [4] conducted their investigation. It was located in the southern boreal subzone of the European part of Russia nearly 12–30 km south of the town Valday. Valday is situated at the federal highway M10, connecting Moscow with St. Petersburg, both being roughly 400 km apart.

Geographically, the investigation site is a part of the Valday Hills on the east European plain. The average annual temperature is +3.2°C, the annual precipitation is 714 mm, and the frost free period covers 128 days [36]. The Valday Hills were formed by late Weichselian end moraines, consisting of hills and many lakes in the depressions. Loamy to clayey as well as sandy sediments constitute the soil development. The latter were chosen for this study. For the chronosequential approach of this study, sites different in self restoration time but comparable in soil texture, climate, and land use history were selected. Subsequent sampling sites were chosen from information obtained from topographic and geological maps, historical literature, and personal communication with local people. Having found adequate sites, five post-agrogenic Podzols profiles of different self restoration ages were dug in September 2007. Frequent Purckhauer drilling ensured that all profiles were representative of the sites. The chronosequential catena covered 3, 20, 55, 100, and 170 years of self restoration. For precise locations and photos of the soil profiles see Kalinina, et al. [4]. The sand percentage of all soils was >85% and the clay fraction ≤ 2%, indicating their pedological similarity. An Albic Podzol, which had never been under agricultural use, was included as a control. Soil morphology was described according to the Russian taxonomy [37] and world reference base (WRB) of soil resources [38]. Bulk and volume samples were taken in triplicate from every horizon; the former ploughed horizons were additionally sampled at 10 cm intervals with increasing depths.

6.2.2 Glomalin-related Soil Protein (GRSP) Extraction

The GRSP was extracted following the method of Wright and Upadhyaya [39], modified by Halvorson and Gonzales [40]. Briefly, 2 g of soil were extracted by a total volume of 50 ml of 50 mM diphosphate (adjusted to pH 8.0) at 128°C during four extraction cycles, each lasting 1 hr. The pooled extract was centrifuged to remove soil particles and protein concentration determined by the Bradford assay [41] using bovine serum albumin (BSA) as a standard.

6.2.3 Density and Particle Size Fractionation

The procedure was conducted in duplicate to obtain free POM and occluded POM of the light fraction (< 1.8 g cm^{-3}) and fraction of particles < 20 μm of the heavy fraction (>1.8 g cm^{-3}) [42]. To avoid slaking, a 30 g air-dried soil sample < 2 mm was capillary-saturated and then immersed in deionized water on a 20 μm mesh sieve for 5 min. The sieve was gently stirred on a rotating shaker (Laborshake Gerhardt Bonn) for 100 revolutions at a frequency of 55 rpm, dried at 40°C, weighed, and the soil material was transferred to a centrifuge tube together with sodium polytungstate at a density of 1.8 gcm^{-3} (Tungsten Compounds (TC), Germany). The filling level was 2/3 of the tube. The tube was hand shaken ten times, and then centrifuged at 1200 g for 10 min. The supernatant was washed with deionized water onto a paper filter (weight band Rotilabo) until the electrical conductivity dropped to < 50 μScm^{-1}. The remaining solution was filtered through a Glass Microfiber filter (GF/A) (referred to as free POM of the light fraction (< 1.8 g cm^{-3})). The residual soil material was dispersed ultrasonically (HF-Generator GM-2200, Sonotrode VS 70T, Bandelin) with energy of 450 Jml^{-1} for complete dispersion of soil micro aggregates [43]. The instrument was calibrated calorimetrically every 4 weeks according to North [44]. The dispersed soil was re-sieved at 20 μm. The material on the sieve was dried and treated to separate the occluded POM, which was not found. The fine fraction < 20 μm was obtained by shaking a horizontal shaker at the beginning of the procedure and the fine fraction < 20 μm gained after re-sieving the ultrasonically dispersed soil was combined and obtained by centrifugation by 4000 g for 18 min. The supernatant was removed and the settled sample was washed intensively with deionized H_2O. The procedure was repeated until the electrical conductivity was < 50 μScm^{-1} and the supernatant was clear. Thereafter, the sample was dried at 40°C and weighed (referred to mineral-associated OC (< 20 μm) of the heavy fraction (>1.8 g cm^{-3}). All fractions were analyzed for total C and N.

The free POM of the light fraction was additionally analyzed by infrared spectroscopy (FTS 7, Fa. Bio-Rad). Duplicate samples of 3–4 mg were ground with 100 mg KBr and pressed (8 t) to a pellet. The subsequent measurements were carried out in the mid-infrared area (225–4000 cm^{-1}). The spectra were transformed into their real OC quantities of total C. The interpretation of the IR spectra was done according to Stevenson [45], Fookeu [46], and Senesi et al. [47].

6.2.4 Grain-size Fractionation

Grain-size fractionation was done according to Kaiser et al. [48]. Destruction of aggregates and sample dispersion was achieved by a two-step ultrasonication (HF-Generator GM-2200, Sonotrode VS 70T, Bandelin). The instrument was calibrated calorimetrically every 4 weeks according to North [44]. The energy used for separating sand particles >200 μm was 60 Jml^{-1}. The energy input for the second dispersion step was 440 Jml^{-1}. Sand and coarse silt particles (>20 μm) were separated by wet-sieving. Medium silt (20–6.3 μm) and fine silt (6.3–2 μm) fractions were recovered by sedimentation. The clay size particles (< 2 μm) were separated by centrifugation at 4400 g for 20 min. Then, the particle-size separates were dried at 105°C and analyzed for total C and N.

The sum of the fractions is lower than 100% because of the specification of the fractionation technique: The OC measurement in many fractions, OC loses, due to sodium polytungstate, and application [32, 49–52].

The C and N contents in dry soil pellets were determined after combustion and spectrometric measurements with a C/N/S analyzer (CHNS-Analyzer Flash EA) within the density and particle size fractions < 20 μm as well as within the grain-size fractions (sand, silt, clay).

Unless stated otherwise, the data were based on three replicates; the standard deviations were always less than 10% of the mean.

6.3 DISCUSSION

This chronosequence showed an enrichment in SOC during self restoration [4], which was caused by the development of an organic surface horizon and not by changes within the mineral horizons (Table 1). In the mineral horizons only slight modifications and peculiarities were observed, including enrichment in the newly developed Ah horizons 3–20 years after self restoration, relatively high C contents within the newly developed albic horizons compared to natural Podzols [20], and relatively constant C contents within relict ploughed horizons. The latter confirms the morphological findings of long existing Ap features and corresponds to the high phosphorus contents even after an abandonment time of 170 years [4] and high root densities. Consequently, the former ploughed horizons act as a favored rooting zone for the newly established vegetation. This assumption was confirmed by the occurrence of high GRSP concentrations. Generally, the SOC contents of the mineral horizons did not change during self restoration, indicating quantitatively balanced carbon dynamics during this process.

TABLE 1 Contents of total OC and C/N ratios, free POM of the density fraction < 1.8 g cm^{-3} and the mineral associated OM (< 20 μm) of the density fraction >1.8 g cm^{-3} of the topsoils after 3, 20, 55, 100, and 170 years under self restoration and of an Albic Podzol, never been cultivated.

Horizon	Depth	C	C/N	Free POM of the density fraction < 1.8 g cm^{-3}			Mineral associated OM (<20 μm) of the density fraction >1.8 g cm^{-3}		
	cm	g kg^{-1}		% of soil	% of SOC	C/N	% of soil	% of SOC	C/N
				3 years					
Ah	10(12)(1)	23.7	15.6	1.33	56.2	12.5	0.62	26.3	12.1
	22	19.6	15.1	1.11	56.6	13.4	0.62	31.7	12.2
Apl	33	19.1	16.5	0.81	42.6	12.7	0.48	25.2	12.7
Ap2	40	8.1	16.6	0.35	42.7	15.6	0.24	30.0	11.8
				20 years					
Ah	2	29.9	14.6	2.25	75.3	14.9	0.61	20.3	13.7
Apl	18	14.9	14.1	0.69	46.1	17.3	0.50	33.5	14.4
Ap2	25(28)	13.7	15.6	0.73	53.2	18.3	0.53	38.6	14.4

TABLE 1 *(Continued)*

Horizon	Depth	C	C/N	Free POM of the density fraction < 1.8 g cm^{-3}			Mineral associated OM (<20 µm) of the density fraction >1.8 g cm^{-3}		
	cm	g kg^{-1}		% of soil	% of SOC	C/N	% of soil	% of SOC	C/N
					55 years				
Ah-E	1	nd	nd	nd	nd	nd	nd	nd	nd
Ap1	8	18.2	27.2	1.13	62.0	32.2	0.54	29.6	18.9
Ap2	19	12.3	27.0	0.68	55.1	33.8	0.39	32.0	18.8
					100 years				
E	2	36.1	28.6	2.46	68.0	27.4	0.86	23.8	21.2
Bsh	5(9)	26.5	26.5	1.08	40.8	30.6	0.93	35.3	22.4
Ap	20	20.0	30.3	1.03	51.3	25.6	0.85	42.4	20.4
					170 years				
E	2	21.7	31.2	1.53	70.5	27.9	0.54	24.9	19.0
Bsh	6	20.6	28.3	1.35	65.6	32.4	0.59	28.8	19.8
Ap	15	15.4	41.8	1.05	68.3	37.6	0.49	31.8	22.8
					Albic Podzol				
E	6(8)	20.9	38.1	1.56	74.5	40.3	0.48	23.1	28.3
Bsh	10	26.4	35.7	1.03	38.9	35.7	0.85	32.2	26.5

nd: not determined
(1) Depths descriptions in parentheses indicate partly thicker horizons.

Occluded POM in the light fraction (< 1.8 g cm^{-3}) was not found, confirming the morphological findings of no aggregates in the studied soils.

With larger amounts in the top than in the subsoils, the highest OC content was found for free POM in the light fraction (< 1.8 g cm^{-3}) and in the sand and coarse silt size separates (Table 1 and 2). Composed of macro OM from plant and animal origin [14], this SOC refers to the active and intermediate pool if charcoal is negligible with a short turnover time from 1 to 100 years [18]. Although a lower amount of active and intermediate OC pools in the arable soil than in the forest soil have been documented in several studies [7, 16, 21–23], we did neither find increasing SOC in the free POM of the light fraction (< 1.8 g cm^{-3}) nor in the sand and coarse silt size separates with increasing duration of self restoration (Table1, 2). We assume that the temporal constant amounts of these fractions in the mineral topsoil during self restoration were caused by compensation of the loss of these fractions from the agronomic phase *via* mineralization and

TABLE 2 Contents of OC within the grain size fractions of the topsoils after 3, 20, 55, 100, and 170 years under self restoration and of an Albic Podzol, never been cultivated.

Horizon	Depth	OC content in grain size					
		Sand		Silt			Clay
		2.0 - 0.2 mm	0.2 - 0.063 mm	0.063 - 0.020 mm	0.020 - 0.0063 mm	0.0063 - 0.002 mm	<0.002 mm
	cm	% of SOC					
		3 years					
Ah	10(12)[1]	33.0	19.3	21.6	8.4	10.8	6.3
Ap1	22	34.6	15.8	22.2	5.7	13.9	7.5
	33	35.0	9.5	16.4	6.9	10.8	8.7
Ap2	40	39.1	6.5	11.1	8.2	14.1	13.1
		20 years					
Ah	2	38.9	12.3	20.5	8.6	10.4	8.3
Ap1	18	20.4	10.5	17.1	12.5	16.6	15.2

TABLE 2 *(Continued)*

Horizon	Depth	OC content in grain size					
		Sand		Silt			Clay
		2.0 - 0.2 mm	0.2 - 0.063 mm	0.063 - 0.020 mm	0.020 - 0.0063 mm	0.0063 - 0.002 mm	<0.002 mm
	cm	% of SOC					
Ap2	25(28)	22.5	9.4	21.2	12.2	21.3	13.1
55 years							
Ah-E	1	nd	nd	nd	nd	nd	nd
Ap1	8	32.2	6.8	10.1	12.3	11.4	9.1
Ap2	19	24.4	7.8	15.2	21.7	13.6	13.3
100 years							
E	2	43.1	3.3	23.9	7.2	13.1	9.1
Bsh	5	41.3	5.1	15.2	15.3	10.9	6.8
Ap	10	37.1	6.5	16.4	23.6	8.9	7.0

TABLE 2 *(Continued)*

Horizon	Depth	OC content in grain size					
		Sand			Silt		Clay
		2.0 - 0.2 mm	0.2 - 0.063 mm	0.063 - 0.020 mm	0.020 - 0.0063 mm	0.0063 - 0.002 mm	<0.002 mm
	cm	% of SOC					
170 years							
E	2	50.1	4.4	14.5	9.0	13.8	8.1
Bsh	6	31.2	4.9	16.2	12.7	14.1	9.3
Ap	15	30.5	9.1	17.3	16.4	10.1	13.8
Albic Podzol							
E	6(8)	41.6	5.2	16.7	13.3	12.6	10.4
Bsh	10	35.4	7.3	8.8	6.3	6.0	6.5

nd: not determined
[1] Depths descriptions in parentheses indicate partly thicker horizons

subsequent humification of newly developing roots. These fractions show an adjusted carbon balance although mineralization and carbon sequestration occurred, confirming other studies [24-26], which state, that the light fraction is not a universal indicator in relation to land use change.

The OC of the heavy fraction (>1.6–2.0 g cm^{-3}) and fine-silt and clay fraction includes organoclay complexes and mineral grains coated with OM, representing a stable SOC or passive pool [14,18] with turnover times of 10–100 years and more for OC in the fine-silt and definitely more than 100 years in the clay fraction. Consequently, the OC of the particles size < 20 μm released from density fraction >1.8 g cm^{-3} and the OC of the fine silt and clay fraction, investigated in this study, compose a stable OC pool with long turnover times. According to John, et al. [16], a decrease of this pool was expected for the studied chronosequence. Nevertheless, the OC ratios of the particles' size < 20 μm in the heavy fractions of the studied topsoils were in the similar range to the OC contents in the fine silt and clay fraction (Table 1 and 2), showing no quantitative alterations of the stable OC pools with increasing duration of self restoration which was unexpected because of increasing podzolization dynamics but indicates once again an adjusted carbon balance.

The C/N ratios of free POM in the light fraction (< 1.8 g cm^{-3}) increased within self restoration (Table 1). These C/N ratios were expected to reach the value of the Albic Podzol (Table 1) but because of particular charcoal enrichment in the latter [4] they did not. The increase of C/N ratios in these fractions resulted from increasing mineralization of agriculture-derived SOM and from the gradual input of qualitatively different organic material with low nitrogen concentration as discussed. The nitrogen being released in mineral topsoil due to SOM mineralization is assimilated by the vegetation, creating a loss for the mineral soil [17]. As plant available phosphorus and potassium showed the same behavior within this chronological sequence [4] an overall shifting of the source of plant nutrients upwards to the forest floor was observed and documented in respect of principle raw humus formation processes by Chertov [27] and Chertov et al. [28].

Although no quantitative modifications of total SOM was observed, the changing C/N ratios of free POM in the light fraction (< 1.8 g cm^{-3}) reflected qualitative alterations during self restoration. This process could not be seen *via* IR spectroscopic analysis (data not shown). The highest and no more increasing C/N values, achieved after 55 years of self restoration (Table 1), indicated a new qualitative OC balance being present after that time. The N content in the sand size fractions of ploughing horizons was under detecting limit (Table 3), indicating partial rotation of an active OC pool already after 20 years of self restoration. Short rotation times were also found for other environmental conditions [16, 29]. For this study, the quick SOM rotation is considered to be due to "fresh" and easy decomposable macro organic residues being characteristic of these OC fractions [14] in combination with accelerated acidification within this chronosequence [4], the latter also discussed by Chertov and Menshikova [30].

TABLE 3 The C/N ratios of the organic matter within the grain size fractions of the topsoils after 3, 20, 55, 100, and 170 years under self-restoration and of an Albic Podzol, never been cultivated.

Horizon	Depth cm	C/N ratios in grain size					
		Sand		Silt			Clay
		Coarse/ Medium	Fine	Coarse	Medium	Fine	
		2.0-0.2	0.2-0.063	0.063-0.020	0.020-0.0063	0.0063-0.002	<0.002
		mm					
		3 years					
Ah	10(12)(1)	31.4	28.8	14.0	12.9	12.0	10.7
Ap1	22	25.0	24.2	14.2	12.1	12.0	11.2
	33	34.7	nd	14.9	14.3	12.0	10.9
Ap2	40	nd	nd	23.2	14.9	12.0	9.6
		20 years					
Ah	2	21.0	24.5	16.5	13.2	12.4	8.3
Ap1	18	nd	nd	19.0	15.5	13.5	9.2
Ap2	25(28)	nd	nd	21.3	18.0	15.5	10.1
		55 years					
Ah-E	1	nd	nd	nd	nd	Nd	nd
Ap1	8	nd	nd	38.9	24.2	23.8	13.2

TABLE 3 *(Continued)*

Horizon	Depth cm	C/N ratios in grain size					
		Sand		Silt			Clay
		Coarse/ Medium	Fine	Coarse	Medium	Fine	
		2.0-0.2	0.2-0.063	0.063-0.020	0.020-0.0063	0.0063-0.002	<0.002
		mm					
Ap2	19	nd	nd	45.2	26.3	24.2	15.0
100 years							
E	2	nd	nd	29.8	30.4	21.7	18.9
Bsh	5	nd	nd	32.0	26.7	21.5	15.1
Ap	10	36.7	nd	30.9	24.0	21.6	13.7
170 years							
E	2	53.1	42.8	24.0	24.4	18.5	14.8
Bsh	6	nd	nd	30.5	28.6	17.1	14.6
Ap	15	nd	nd	46.9	29.9	24.9	17.5
Albic Podzol							
E	6(8)	nd	nd	44.6	38.0	28.5	17.7
Bsh	10	nd	nd	31.6	26.8	22.9	17.5

nd - not determined

[1] Depths descriptions in parentheses indicate partly thicker horizons

The qualitative changes in SOM within self restoration were also found by increasing C/N ratios for OM of the particles' size < 20 μm in the heavy fraction (>1.8 gcm^{-3}) (Table 1) and for OM of the fine-silt and clay fraction (Table 3). Lower C/N ratios with decreasing grain sizes indicate less changes of SOM in the same order. These results confirm that SOM changes also affect the stable OC pool, also reported by others [16, 23, 29, 31], showing mean ages of stable SOC in the range from 50 to 260 years. According to the C/N ratios in the heavy fraction and fine-silt clay fractions, the first clear qualitative alterations of the stable OC pool found in this study occurred after 55 years of self restoration. This alteration might be due to portions of rapid cycling C, being among the heavy fraction with long turnover times according to Golchin, et al. and Swanston, et al. [32-34], in combination with quick podzolization found for this chronosequence [4], as already discussed. Supply might be induced by highly decomposed OM from root litter and by leaching of highly oxidized water-soluble organic products from the forest floor into the mineral soil horizons and stabilization therein by association with mineral phases [35]. Although the initiation was quick, "a complete new qualitative balance" of the stable SOM was not achieved within 170 years of self restoration. This is shown by the C/N ratios of OM of the particles' size < 20 μm in the heavy fraction (>1.8 g cm^{-3}) and in the fine-silt and clay fractions at the end of the chronosequence (Table 1 and 3), being lower than those of the Albic Podzol.

6.4 RESULTS

The total SOC contents stayed relatively constant during self restoration (Table 1). This was also obvious for the relictic ploughed horizons, showing a mean of 17.6 g kg^{-1} for the 3 years old soil under self restoration and 15.4 g kg^{-1} for the 170 years old soil. Slight SOC enrichments were found in the newly developed Ah horizons 3–20 years after self restoration with 23.7 and 29.9 g kg^{-1} respectively. The SOC contents of the newly developed albic horizons were 36.1 and 21.7 after 100 and 170 years of self restoration, whereas it was 20.9 in the same horizon of the Albic Podzol.

To test the occurrence of young living roots and to exclude the residual character from former land use GRSP concentration was measured. As Glomalin is a protein produced by arbuscular mycorrhizal fungi during colonization, Glomalin concentration is correlated to the abundance of metabolizing roots [19]. The GRSP concentration was 2.1–2.7 mg g^{-1} in the former ploughed horizons of soils being 55, 100, and 170 years under self restoration whereas it was 1.9–3.2 mg g^{-1} in the Ah, Ap horizons of the 3 and 20 years' soils under self restoration.

The OC content of free POM in the light fraction (< 1.8 g cm^{-3}) was 38.9–75.3% of SOC (Table 1). The OC ratios of the particles' size < 20 μm in the heavy fractions of the studied topsoils were in the range of 20.3–42.4% of SOC. Therefore, neither fraction showed a significant change during self restoration. Occluded POM in the light fraction (< 1.8 g cm^{-3}) was not found. The C/N ratios of free POM in the light fraction (< 1.8 g cm^{-3}) increased under self restoration: 13.8–16.8–33.0–27.9–32.6 (in the mean) after 3, 20, 55, 100, and 170 years of abandonment. The C/N ratio of the free POM in the light fraction (<1.8 g cm^{-3}) of the Albic Podzol was with 36 in the mean remarkably high. The C/N ratios of OM of the particles size < 20 μm in the heavy fraction (>1.8 g cm^{-3}) also increased within self restoration: 12.2–14.2–18.9–21.3 –20.5

(in the mean) after 3, 20, 55, 100, and 170 years of abandonment. The Albic Podzol showed a C/N ratio 27.4. The IR spectroscopic analysis of the OM in the light fractions did not show differences within self restoration in respect to the composition of the functional groups as well as to their relative rates (data not shown).

The OC content in the sand and coarse silt size separates was 47.4–73.9% of SOC (Table 2). The OC contents in the fine silt and clay fraction were in a range of 12.5–34.4% of SOC. As a result, both fractions did not show a significant change during self restoration. The N contents in the sand size fractions of ploughing horizons were under detecting limit from 3 years of self restoration and thereafter (Table 3). The C/N ratios of OM of the fine-silt and clay fraction also increased within self restoration but with less extent, showing C/N ratios of 12 (mean values) at the beginning and 20 at the end of the chronosequntial observation for the silt and 11 and 15 for the clay fraction, respectively.

6.5 CONCLUSION

Although the sandy post-agrogenic soils in the southern Taiga sub-zone of European Russia indicated a carbon sink during self restoration by increasing carbon sequestration within the developing organic surface layers, the mineral horizons showed only small alterations in total OC contents and no quantitative but qualitative modifications in respect of different functionally OC pools. The constant quantities are supposed to result from simultaneously running supply and mineralization processes. The changing qualities indicate the formation of "a new balance" of active SOM within some decades, whereas this process is not yet completed for the passive SOM after 170 years of self restoration. Also expressed by remaining soil morphological ploughing features, this study confirms that carbon dynamics are not balanced after 170 years of self restoration. Soil recovery from agriculture is a long-term or infinite process.

KEYWORDS

- **Bovine serum albumin**
- **Morphological features**
- **Organic matter**
- **Podzols**
- **Soil organic matter**

AUTHORS' CONTRIBUTIONS

Olga Kalinina carried out literature research, participated in soil sampling, performed soil analysis, and drafted the manuscript. Sergey V. Goryachkin and, Dmitriy I. Lyuri, conceived of the project idea, carried out the study of historical land use, participated in soil sampling. Nina A. Karavaeva helped in characterisation of the soils. Luise Giani is participated in coordination of the study and its design, and drafted the manuscript. All authors read and approved the final manuscript.

ACKNOWLEDGMENT

This work was supported by the Deutsche Forschungsgemeinschaft by grant GI 17120-1. We thank Marie Spohn for the analysis of Glomalin, Gunda Sängerlaub, and Brigitte Rieger for their technical assistance.

REFERENCES

1. Lyuri, D. I., Karavaeva, N. A., Nefedova, T. G., Konyushkov, B. D., and Goryachkin, S. V. *Self Restoration of Post-Agrogenic Soils: Recent Process of Late Antropocene. Abstracts of 18th World* Congress of Soil Science. Session 1.0 B Soil Change in Anthropocence, pp. 27–2 (2006).
2. Lyuri, D. I., Goryachkin, S. V., Karavaeva, N. A., Nefedova, T. G., and Denisenko, E. A. Regularities of agricultural land withdrawal in Russia and in the world and processes of post-agrogenic development of fallows. In *Agroecological status and prospects of use of lands of Russia withdrawn from active agricultural rotation* (in Russian). A. L. Ivanov (Ed.), Dokuchaev Soil Science Institute, Moscow, pp. 45–71 (2008).
3. Ramankutty, N. *Global land-cover change: Recent progress, remaining challenges*. In: Land use and land-cover change. E. F. Lambin and H. J. Geist (Eds.). Springer, Berlin, pp. 9–41 (2006).
4. Kalinina, O., Goryachkin, S. V., Karavaeva, N. A., Lyuri, D. I., Najdenko, L., and Giani, L. Self restoration of post-agrogenic sandy soils in the southern taiga of Russia: Soil development, nutrient status, and carbon dynamics. *Geoderma*, **52**, 35–42 (2009).
5. Kalinina, O. and Giani, L. Degradation von Plaggeneschen und Konsequenzen für ihre ökologische Bewertung. *Mitteilungen Deutsche Bodenkundliche Gesellschaft*, **108**, 105–106 (2006).
6. Lüst, C. and Giani, L. Merkmale von Böden unter rezenten Wäldern, die auf ehemals landwirtschaftlich genutzten Flächen stocken. *Drosera*, **1/2**, 27–34 (2006).
7. Degryze, S., Six, J., Paustian, K., Morris, S. J., Paul, E. A., and Merckx, R. Soil organic carbon pool changes following land use conversions. *Global Change Biology*, **10**, 1120–1132 (2004).
8. Hooker, T. D. and Compton, J. E. Forest ecosystem carbon and nitrogen accumulation during the first century after agricultural abandonment. *Ecological Applications*, **13**(2), 299–313 (2003).
9. Post, W. M. and Known, K. C. Soil carbon sequestration and land use change: processes and potential. *Global Change Biology*, **6**, 317–327 (2000).
10. Richter, D. D., Markewitz, D., Trumbore, S. E., and Wells, C. G. Rapid accumulation and turnover of soil carbon in a re-establishing forest. *Nature*, **400**, 56–58 (1999).
11. Thuille, A. and Schulze, E. D. Carbon dynamic in successional and afforested spruce stando in Thuringia and the Alps. *Global Change Biology*, **12**, 325–342 (2006).
12. Paul, K. I., Polglase, P. J., Nyakuengama, J. G., and Khanna, P. K. Change in soil carbon following afforestation. *Forest Ecology and Management*, **168**, 241–257 (2002).
13. Wilde, S. A., Iyer, J. G., Tanzer, C., Trautmann, W. L., and Watterston, K. G. *Growth of Wisconsin coniferous plantations in relation to soils*. University of Wisconsin, Madison, Research Bulletin, **262**, 1–81 (1965).
14. Christensen, B. T. Physical fractionation of soil and organic matter in primary particle size and density separates. In: *Advances in soil science*. B. A. Stewart (Ed.) Springer-Verlag, New York, pp. 1–90 (1992).
15. Flessa, H., Amelung, W., Helfrich, M., Wiesenberg, G. L. B., Gleixner, G., Brodowski, S., Rethemeyer, J., Kramer, C., and Grootes, P. M. Storage and stability of organic matter and fossil carbon in a Luvisol and Phaeozem with continuous maize cropping: A synthesis. *Journal of Plant Nurtrition and Soil Science*, **171**, 36–51 (2008).
16. John, B., Yamashita, T., Ludwig, B., and Flessa, H. Storage of organic carbon in aggregate and density fractions of silty soils under different types of land use. *Geoderma*, **128**, 63–79 (2005).
17. Chertov, O. G. and Komarov, A. S. A model of soil organic matter dynamics. *Ecological Modelling*, **94**, 177–189 (1997).

18. Von Lützow, M., Kögel-Knabner, I., Ludwig, B., Matzner, E., Flessa, H., Ekschmitt, K., Guggenberger, G., Marschner, B., and Kalbitz, K. Stabilization mechanisms of organic matter in four temperate soils: Development and application of a conceptual model. *Journal of Plant Nutrition and Soil Science*, **171**, 111–124 (2008).
19. Wright, S. F., Franke-Snyder, M., Morton, J. B., and Upadhyaya, A. Time-course study and partial characterization of a protein on hyphae of arbuscular mycorrhizal fungi during active colonization of roots. *Plant and Soil*, **181**, 193–203 (1996).
20. Sauer, D., Sponagel, M., Giani, L., Jahn, R., and Stahr, K. Podzol: Soil of the year 2007. A review on its genesis, occurrence, and functions. *Journal of Plant Nutrition and Soil Science*, **170**, 581–597 (2007).
21. Mendham, D. S., Heagney, E. C., Corbeels, M., O'Connell, A. M., Grove, T. S., and McMurtrie, R. E. Soil particulate organic matter effects on nitrogen availability after afforestation with Eucalyptus globules. *Soil Biology and Biochemistry*, **36**(7), 1067–1074 (2004).
22. Six, J., Callewaert, S., Lenders, S., De Gryze, S., Morris, S. J., Gregorich, E. G., Paul, E. A., and Paustian, K. Measuring and understanding carbon storage in afforested soils by physical fractionation. *Soil Science Society of America Journal*, **66**, 1981–1987 (2002).
23. Yamashita, T., Flessa, H., John B., Helfrich, M., and Ludwig, B. Organic matter in density fractions of water-stable aggregates in silty soils: Effect of land use. *Soil Biology and Biochemistry*, **38**(11), 3222–3234 (2006).
24. Billings, S. A. Soil organic matter dynamics and land use change at a grassland/forest ecotone. *Soil Biology and Biochemistry*, **38**(9), 2934–2943 (2006).
25. Don, A., Scholten, T., and Schulze, E. D. Conversion of cropland into grassland: Implications for soil organic carbon stocks in two soils with different texture. *Journal of Plant Nutrition and Soil Science*, **172**, 53–62 (2009).
26. Lima, A. M. N., Silva, I. R., Neves, J. C. L., Novais, R. F., Barros, N. F., Mendonca, E. S., Smyth, T., Moreira, M. S., and Leite, F. P. Soil organic carbon dynamics following afforestation of degraded pastures with eucalyptus in southeastern Brazil. *Forest Ecology and Management*, **235**(13), 219–231 (2006).
27. Chertov, O. G. *Ecology of Forest Lands* (in Russian). Soil-Ecological Studies of Forest Sites. Nauka (Science Publication), Moscow (1981).
28. Chertov, O. G., Komarov, A. S., Nadporozhskaya, M. A., Bykhovets, S. A., and Zudin, S. L. Simulation study of nitrogen supply in boreal forests using model of soil organic matter dynamics ROMUL. In: *Plant Nutrition—Food Security and Sustainability of Agro-Ecosystems*. W. J Horst., et al (Eds.). Kluwer, Dordrecht, Boston, and London, pp. 900–901 (2001).
29. Golchin, A., Oades, J. M., Skjemstad, L. O., and Clarke, P. 13C NMR-spectroscopy in density fractions of an oxisol under forest and pasture. *Australian Journal of Soil Research*, **33**(1), 59–76 (1995).
30. Chertov, O. G. and Menshikova, G. P. About influence of acid precipitation on the forest soils. *USSR and Isvestiya* (in Russian), **6**, 906–913 (1983).
31. Bird, J. B., Kleber, M., and Torn, M. S. 13C and 15N stabilization dynamics in soil organic matter fractions during needle and fine root decomposition. *Organic Geochemistry*, **39**, 465–477 (2008).
32. Golchin, A., Oades, J. M., Skjemstad, L. O., and Clarke, P. Soil structure and carbon cycling. *Australian Journal of Soil Research*, **32**(5), 1043–1068 (1994).
33. Swanston, C. W., Caldwell, B. A., Homann, P. S., Ganio, L., and Sollins, P. Carbon dynamics during a long-term incubation of separate and recombined density fractions from seven forest soils. *Soil Biology and Biochemistry*, **34**, 1121–1130 (2002).
34. Swanston, C. W., Torn, M. S., Hanson, P. J., Southon, J. R., Garten, C. T., Hanlon, E. M., and Ganio, L. Initial characterization of processes of soil carbon stabilization using forest stand-level radiocarbon enrichment. *Geoderma*, **128**, 52–62 (2005).
35. Cerli, C., Celi, L., Kaiser, K., Guggenberger, G., Johansson, M. B., Cagnetti, A., and Zanini, E. Changes in humic substances along an age sequence of Norway spruce stands planted on former agricultural land. *Organic Geochemistry*, **39**(9), 1269–1280 (2008).

36. *National Parks of Russia* (in Russian). I. V. Chebakova (Ed.). Biodiversity Conservation Centre, Moscow (1996).
37. Shishov, L. L., Tonkonogov, W. D., Lebedeva, J. J., and Gerasimova, M. J. *Russian soil classification* (in Russian) Oykumena, Smolensk (2004).
38. IUSS Working Group WRB: WRB 2006. World Soil Res. Repts. 103. FAO, Rome (2006).
39. Wright, S. F. and Upadhyaya, A. A survey of soils for aggregate stability and glomalin, a glycoprotein produced by hyphae of arbuscular mycorrhizal fungi. *Plant and Soil*, **198**, 97–107 (1998).
40. Halvorson, J. J. and Gonzales, J. M. Bradford reactive soil protein in Appalachian soils: distribution and response to incubation, extraction reagent and tannins. *Plant and Soil*, **286**, 339–356 (2006).
41. Bradford, M. M. A rapid and sensitive for the quantitation of microgram quantities of protein utilizing the principle of protein-dye binding. *Analytical Biochemistry*, **72**, 248–254 (1976).
42. Leifeld, J. and Kögel-Knabner, I. Soil carbon matter fractions as early indicators for carbon stock changes under different land-use?, *Geoderma*, **124**, 143–155 (2005).
43. Schmidt, M. W. I., Rumpel, C., and Kögel-Knabner, I. Evaluation of an ultrasonic dispersion procedure to isolate primary organomineral complexes from soils. *European Journal of Soil Science*, **50**, 87–94 (1999).
44. North, P. F. Towards an absolute measurement of soil structural stability using ultrasound. *Journal of Soil Science*, **27**, 451–459 (1976).
45. Stevenson, F. J. *Humus chemistry: genesis, composition, reactions*. Wiley, New York (1982).
46. Fookeu, U. *Huminsären in Oberflächensedimenten der Nordsee*: *Indikatoren für terrestrischen Eintrag*? Ph.D. Thesis. University Oldenburg, Germany (2000).
47. Senesi, N., D'Orazio, V., and Ricca, G. Humic acids in the first generation of EUROSOILS. *Geoderma*, **116**, 325–344 (2003).
48. Kaiser, K., Eusterhues, K., Rumpel, C., Guggenberger, G., and Kögel-Knabner, I. Stabilization of organic matter by soil minerals—investigations of density and particle-size fractions from two acid forest soils. *Journal of Plant Nutrition and Soil Science*, **165**, 451–459 (2002).
49. Chenu, C. and Plante, A. F. Clay-sized organo-mineral complexes in a cultivation chronosequence: revisiting the concept of the "primary organo-mineral complex". *European Journal of Soil Science*, **57**, 569–607 (2006).
50. Magid, J., Gorissen, A., and Giller, K. E. In search of the elusive "active" fraction of soil organic matter: three size-density fractionation methods for tracing the fate of homogeneously C-14-labelled plant materials. *Soil Biology and Biochemistry*, **28**, 89–99 (1996).
51. Rovira, P., Casals, P., Romanya, J., Bottner, P., Couteaux, M., and Vallejo, R. Recovery of fresh debris of different sizes in density fractions of two contrasting soils. *European Journal of Soil Biology*, **34**, 31–37 (1998).
52. Rovira, P. and Ramon Vallejo, V. Physical protection and biochemical qualità of organic matter in Mediterranean calcareous forest soils: a density fractionation approach. *Soil Biology and Biochemistry*, **35**, 245–261 (2003).

PART II
SOIL BIOTIC INTERACTIONS

7 Tropical Earthworms

Radha D. Kale and Natchimuthu Karmegam

CONTENTS

7.1 INTRODUCTION

This chapter highlights the research carried out by different scientists in India on aspects of earthworm population dynamics and species diversity, associated with other soil fauna and microflora. It also deals with the importance of earthworm activity on physicochemical properties of soil with reference to India and other tropical countries. Stress is laid on the earthworm plant association and importance of the secretions of earthworms as plant growth stimulators. Moreover, the earthworm species reported and being utilized for vermicomposting in India are discussed, since vermicomposting is the ultimate technology that renders for the improvement of soil fertility status and plant growth. Earthworms serve as indicators of soil status such as the level of contamination of pollutants: agrochemicals, heavy metals, toxic substances, and industrial effluents; human-induced activities: land management practices and forest degradation. In all these fields there is lacuna with respect to contributions from India when compared to the available information from other tropical countries. There is lot of scope in the field of research on earthworms to unravel the importance of these major soil macrofauna from holistic ecological studies to the molecular level.

Earthworms belonging to Phylum Annelida, Class Chaetopoda, and Order Oligochaeta occupy a unique position in animal kingdom. They are the first group of multicellular, eucoelomate invertebrates who have succeeded to inhabit terrestrial environment. They form major soil macrofauna. Their species richness, abundance, and distribution pattern reflect on edaphic and climatic factors of the geographical zone. They serve as "bioindicators" to understand the physicochemical characteristics of their habitat. Their horizontal and vertical stratification and abundance contribute to pedogenesis and soil profile. Encouraging their establishment through no tillage or shallow ploughing and enriching soil with organic matter incorporation has resulted in improving soil fertility. This has been experimented for several decades at Rothamsted Research Station, U.K. The interaction of earthworms and other microflora and fauna has given much scope for understanding of soil community and its influence on ground primary production.

Distinctive habitat, food niches, and adaptive mechanisms of earthworms have opened up new fields for investigations on their role in organic waste management. One of the advantageous factors in this field is the use of earthworms to minimize the degradable organic matter and to use the same as bioresource for organic manure production. The manure produced serves as good source of soil amendment. The ecologically distinguished epigeic earthworms are used for producing the organic manure, "vermicompost." This has gained attention of garden lovers, agriculturists, and agro industries to convert organic matter generated at different levels into rich, odorless, and free flowing compost to support sustainable agriculture.

7.2 EARTHWORMS: COMPONENTS OF SOIL BIOTA

Earthworms form one of the major macrofauna among soil biota to maintain dynamic equilibrium and regulate soil fertility. Their existence depends on adequate moisture, soil texture, pH, electrolyte concentration, and food source in the ecosystem. This clearly indicates the interdependency of the environmental factors to the survival of earthworms; when such conditions are created, they further contribute to soil fertility through their activity.

7.3 FOOD NICHES OF EARTHWORMS

Degradation of leaf material commences from the time it detaches itself from the plant and drops to ground to add to litter. Earthworms are the major secondary decomposers in the soil faunal community. They feed on decomposed organic material at different levels of degradation. Lee [1] has suggested that earthworms survive on microorganisms, micro and mesofauna associated with ingested dead tissue. According to Lee, earthworms that feed near the surface on decomposing litter and at the root zone on dead roots are the detritivores and those remain at subsurface and consume large quantities of soil are geophagous earthworms.

Lavelle [2] has categorized geophagous earthworms as polyhumic, oligohumic, and mesohumic based on the proportion of humus and soil in their feed. Through factorial analysis, Lavelle has given the explanation that temperature differences with latitude and litter characteristics like quantity and decomposability determine the variations observed with reference to their distribution. The detritivorous epigeic earthworms form the major component of earthworm fauna in temperate regions and mesohumic endogeic earthworms are predominant in tropical forests. There is minimum representation of mesohumic earthworms in temperate regions. Oligohumic earthworms that feed on soil having very low level of organic matter are abundant only in tropical regions.

Lavelle [2] considers polyhumic earthworms as more stable fraction of earthworm community occupying different soil strata as topsoil feeders to species of rhizosphere in tropical regions. Thus, tropical earthworms depend more on soil mixed with different levels of humic substances rather than surface litter. More stable environments like heavy RF areas (2000 to above 4000 mm rain/annum) in the state of Karnataka, India, have greater diversity of earthworms than the dry areas (< 600–900 mm rain/annum). The geophagous earthworms of mesohumic and polyhumic types are widely distributed in places receiving heavy RF in this subtropical part of the country (Tables 1 and 2).

TABLE 1 Earthworm distribution in Southern Karnataka (India) in different agroclimatic zones including coastal plains, hilly regions, and interior plains.

S. No.	Species	Moisture Level	Soil Type	Vertical Distribution (cm)	Food Niche	Population Density (no/100 m^2)
1	*Curgeona narayani*	Wet land-in waterlogged soil	Red loamy soil	Up to 45	Mesohumic	640–11,250
2	*Dichogaster affinis*	20–40	Red loamy, alluvial and lateritic	5–10	Mesohumic to polyhumic	60–250
3	*D. bolaui*	20–40	"	''	''	60–450
4	*D. curgensis*	20–40	Red loamy	''	Polyhumic	25–200
5	*D. modigliani*	20–40	Red sandy	''	Mesohumic	10–25
6	*D. saliens*	20–40	Red sandy	''	''	65–265
7	*Drawida ampullacea*	> 40	Red loamy	10–20	Polyhumic	275–930
8	*D. barwelli*	> 50	Red loamy to sandy soil	10–30	''	275–576
9	*D. barwelli impertusa*	> 50	Red loamy	''	''	120–430
10	*D. calebi*	> 50	Red loamy to sandy soil	10–30	Polyhumic	80–1200
11	*D. ferina*	40–50	Red loamy	20–30	Mesohumic	40–340
12	*D. ghatensis*	40–50	''	10–20	''	450–1350
13	*D. kanarensis*	40–50	''	''	''	85–400

TABLE 1 *(Continued)*

S. No.	Species	Moisture Level	Soil Type	Vertical Distribution (cm)	Food Niche	Population Density (no/100 m²)
14	*D. lennora*	40–50	Red sandy soil	ʼʼ	ʼʼ	15–30
15	*D. modesta*	40–50	ʼʼ	10–30	ʼʼ	4–500
16	*D. paradoxa*	> 40	Red loamy to alluvial	10–20	Polyhumic	1700–2500
17	*D. pellucida pallida*	> 40	Lateritic to Red loamy	ʼʼ	Mesohumic	4–500
18	*D. scandens*	> 40	Red sandy loam	5–10	Polyhumic	10–350
19	*D. sulcata*	> 40	Alluvial soil	10–30	Polyhumic	65–235
20	*Glyphidrillus annandalei*	> 40	Sandy bed to Red loam	20–45	Oligohumic	130–1600
21	*Gordiodrilus elegans*	> 40	Red sandy loam	10–40	Mesohumic	24–200
22	*Hoplochaetella kempi*	3040	Lateritic to alluvial	10–30	Polyhumic	10–430
23	*H. suctoria*	30–40	Alluvial	10–20	ʼʼ	50–240
24	*Hoplochaetella sp.*	40–50	Red loam	20–40	ʼʼ	460–3330
25	*Howascolex sp.*	30–40	Red loam	10–30	ʼʼ	145–2500
26	*Lampito mauritii*	20–30	Red sandy to lateritic	10–30	Mesohumic	720–2190
27	*Mallehula indica*	30–40	Red loam	10–20	Mesohumic	180–880
28	*Megascolex filiciseta*	30–40	Lateritic	5–10	Polyhumic	15–330
29	*M. insignis*	30–40	Alluvial	5–20	Polyhumic	65–800
30	*M. lawsoni*	30–40	Red loam to sandy loam	10–30	Mesohumic	120–1000
31	*M. konkanensis*	30–40	Lateritic to alluvial	20–45	Mesohumic	20–3900
32	*Metaphire houlleti*	> 40	Alluvial and Red loam	10–40	Polyhumic	18–2140

TABLE 1 *(Continued)*

S. No.	Species	Moisture Level	Soil Type	Vertical Distribution (cm)	Food Niche	Population Density (no/100 m^2)
33	*Octochaetona albida*	30–40	Red loam	10–20	Polyhumic	150–650
34	*O. Beatrix*	20–30	Sandy loam	' '	' '	40–335
35	*O. rosea*	30–40	Alluvial	10–20	Mesohumic	15–120
36	*P. excavatus*	> 40	Organic layer	0–5	Detritivore	18–8000
37	*Plutellus timidus*	30–40	Alluvial	10–15	Mesohumic	60–460
38	*Polypheretima elongata*	> 40	Sandy loam to Red loam	30–60	Mesohumic	194–4000
39	*Pontoscolex corethrurus*	30–50	Sandy, alluvial, loamy, lateritic	5–15	Mesohumic to polyhumic	250–7100

The acceptance level of various leaf litters shows positive correlation to nitrogen and carbohydrate contents and negative correlation to polyphenol content [3]. Ganihar [4] studied the litter feeding of *Pontoscolex corethrurus* in a multiple choice test. He found variations in degree of acceptability of different litter that showed positive correlations to levels of OC and nitrogen content. The least preference for *Eucalyptus camaldulensis* and *Acacia auriculiformis* was linked with high levels of polyphenols. It has been shown that *Lampito mauritii* exhibited similar preference either for partially decomposed large pieces of leaf material of different types or for powdered leaves mixed with agar base [5]. It could be inferred that apart from physical nature of leaf matter, chemical compounds in them serve as attractants or repellants (Tables 3 and 4). Ganihar [4] is of the view that in land reclamation sites, if earthworms have to be introduced, it is essential to develop ground plant community. Litter from such plants when mixes with soil, at different levels of decomposition, serves as feed to developing earthworm population. The available carbon source encourages population growth of earthworms [6]. In India, *Lampito mauritii* is the most widely distributed earthworm in different agroecosystems [7–12]. This earthworm preferred decomposing grass of paddy (*Oryza sativa*) and finger millet (*Eleucine coracana*) to other leaf litter [5]. The grasses when developed in reclamation sites can form an ideal base for establishment of *Lampito mauritii* to bring about improvement in soil structure and finally chemical and biological activities. Food preference and sensitivity to other edaphic factors determine the possibility of introduction of earthworms for land reclamation.

TABLE 2 Habitat preference of widely distributed earthworm species *Lampito mauritii* and *Pontoscolex corethrurus* at study sites.

District	Agroclimatic zone	Mean annual RF (mm)	Soil type	Earthworm species	Habitat preference
Bangalore	Eastern dry zone	700–900	Red loamy soil	Lampito mauritii	Arable lands
Kolar	,,	600–800	Lateritic and red sandy soil	,,	Grasslands
Tumkur	,,	,,	Red sandy soil	,,	Grasslands and arable lands
Chickmagalur	South transition zone	900–1000	Red loamy soil	Species richness than species dominance	Varied habitats
Chickmagalur	Hilly zone	2000–3000	Red loamy soil	Pontoscolex corethrurus	Plantations
Coorg	Hilly zone	2000–>4000	,,	,,	Grasslands and plantations
South Kanara	Coastal zone	3000 >4000	Coastal alluvial soil	P. corethrurus and Megascolex konkanensis	Grasslands, plantations, arable lands

Table 3. Disintegration of different leaf matters due to selective feeding by earthworm *Lampito mauritii* [5].

Leaf matter	1	2	3	4	5	6	7	8	9
Millet straw	70.00	50.00	55.00	—	—	—	—	—	—
Paddy straw	48.00	11.00	27.50	22.00	13.00	33.00	—	—	—
Cashew litter	—	—	—	38.00	24.00	39.00	22.50	2.60	67.00
Mango litter	—	—	—	48.00	30.00	50.00	30.70	28.60	6.30
Guava litter	—	—	—	44.00	14.00	83.00	25.00	23.00	—
Eucalyptus litter	—	—	—	32.00	10.00	61.00	31.00	24.40	11.60

Note: Column: 1–3 data for 1st month (1) Percent loss of litter per month due to microbial degradation and feeding by earthworms. (2) Percent microbial degradation per month. (3) Rate of litter consumption (mg) for hundred earthworms per day. Column: 4–6 data for IInd month (4) Percent loss of litter per month due to microbial degradation and feeding by earthworms. (5) Percent microbial degradation per month. (6) Rate of litter consumption (mg) for hundred earthworms per day. Column: 7–9 data for IIIrd month (7) Percent loss of litter per month due to microbial degradation and feeding by earthworms. (8) Percent microbial degradation per month. (9) Rate of litter consumption (mg) for hundred earthworms per day. The table also shows the acceleration of litter breakdown in presence of earthworms.

TABLE 4 Artificial diet (1 : 8 by weight) of agar and different leaf litter powder on feeding of earthworm Lampito mauritii in relation to C/N of diets [5].

Litter powder in agar	Daily food intake mg/day/adult	C/N of the feed
Paddy straw	8.05 ± 0.28	37
Millet straw	7.07 ± 1.23	45
Mango litter	8.67 ± 1.27	19
Guava litter	3.25 ± 0.79	45
Cashew litter	4.44 ± 1.10	30
Eucalyptus litter	1.62 ± 0.59	42
Agar only (control)	2.53 ± 1.23	38

Number of observations = 3; Palatability depends on texture as well as chemical nature of the feed.

7.4 EARTHWORM ACTIVITY ON PHYSICOCHEMICAL PROPERTIES OF SOIL

Earthworms are the major macrofauna in the soil community. They are distributed at different depths in soil strata. The litter feeders, which are not burrowers, constitute a very small number in tropical situations. The burrowing endogeic earthworms live in horizontal and vertical burrows constructed in soil strata. They make these burrows partly by ingesting soil particles through their way and partly by pushing the soil to the sides [13]. The ingested soil along with organic matter passes through the gut and undigested matter is released at the opening of the burrow on soil surface or at the subsurface as castings. The subsurface castings contribute to soil profile [1].

The burrows of earthworms, which run horizontally or vertically depending on burrow forming ability of species, will determine the possible physical effects on soil characteristics. In temperate regions where deep burrowing anecic earthworms are of common occurrence, it is opined that infiltrations can bring about leaching of nutrients from soils to ground water. The leachate volume may show an increase of 412 folds due to their activity [14]. Introduction of *Aporrectodea caliginosa* into coniferous forest soils resulted in fiftyfold increase in concentration of nitrate and cations in soil solution. But the amount that entered ground water or plant system remained undetermined [15]. One of the major contributions of burrowing activity of earthworms is in affecting soil porosity [16, 17]. The major impact on hydrology has been worked out with respect to activity of anecic earthworm *Lumbricus terrestris* [18]. Information is lacking in India with respect to burrows of earthworms, their structure, and any variations observed depending on soil type. Influence of organic matter, agricultural practices on earthworm population, and similarly the role of earthworms in modifying the situations in cultivable lands are very meager in a country having diversity and abundance of the populations in different agroecosystems. Reddy et al. [19] reported the influence of various management practices affecting density and surface cast production. The casts of the earthworm, *Pontoscolex corethrurus*, and the surrounding soil in an undisturbed forestfloor in Sirumalai Hills, Tamil Nadu (South India) showed that the percentage of moisture content, organic carbon, and total nitrogen in the worm casts were higher and significantly differed from the values obtained in the surrounding soil [20].

According to the recent report by Julka et al. [21], in India, there are 590 species of earthworms with different ecological preferences but the functional role of the majority of the species and their influence on the habitat are lacking. Recently Karmegam and Daniel [11] reported the correlation of soil and environmental parameters on the abundance of ten different earthworm species belonging to four families, namely, Megascolecidae (*Lampito mauritii, L. kumiliensis*, and *Megascolex insignis*), Octochaetidae (*Dichogaster bolaui, D. saliens*, and *Octochaetona thurstoni*), Moniligastridae (*Drawida chlorina, D. paradoxa*, and *D. pellucida pallida*), and Glossoscolecidae (*Pontoscolex corethrurus*) in the study that was carried out at different locations in Dindigul District (South India). The fluctuations in populations of earthworms were observed during the monthly collections in course of 3 years in all the selected sites. In the survey carried out from 1997 to 1999, the predominant species that were recorded as maximum number of earthworms/m^2 in sites 110 were *D. pellucida pallida* (Jan. 1998-70.44), *D. pellucida pallida* (Dec. 1999-32.30), *L. mauritii* (Feb. 1998-55.22), *D. pellucida pallida* (Dec. 1999-25.54), *L. mauritii* (Dec. 1997-66.78), *L. mauritii* (Nov. 1997-43.40), *L. mauritii* (Jan. 1999-44.60), *P. corethrurus* (Nov. 1997-58.34), *P. corethrurus* (Dec. 1999-64.30), and *P. corethrurus* (Dec. 1998-107.60) [22].

The biomass dynamics also showed wide fluctuation among the species in relation to the months of collection from different collection sites. The highest worm biomass was recorded during December to February and certain species were totally absent during certain periods of the survey. The total biomass of different species recorded in the monthly observation over a period of 3 years (1997–1999) varied in various study sites. The highest biomass of the respective earthworm species as well as the month and year of its occurrence in the study sites 110 as recorded includes *D. pellucida pallida* (30.63 g/m^2 during Feb. 1998), *D. pellucida pallida* (22.88 g/m^2 during Jan. 1998), *D. pellucida pallida* (29.27 g/m^2 during Dec. 1999), *D. pellucida pallida* (20.20 g/m^2 during Dec. 1999), *D. pellucida pallida* (44.65 g/m^2 during Dec. 1999), D. pellucida pallida (22.38 g/m^2 during Dec. 1999), *D. pellucida pallida* (29.66 g/m^2 during Jan. 1998), *P. corethrurus* (15.20 g/m^2 during Dec. 1998), *D. bolaui* (19.79 g/m^2 during Jan. 1999), and *P. corethrurus* (26.34 g/m^2 during Dec. 1998), respectively [22]. Among the earthworm species studied, *L. kumiliensis* has been reported for the first time in Sirumalai Hills of Tamil Nadu, India [23]. This is the only study to highlight the cyclic fluctuations in the earthworm populations for a continuous period of 3 years and variations in the species structure at different time intervals. Still the information on the physicochemical changes in the soil with respect to species composition at given time is not clear. A composite study on microbial association with the predominant earthworm species at a given time may provide necessary information on its ecological role.

7.5 FACTORS INFLUENCING THE ABUNDANCE OF EARTHWORM POPULATIONS

The percentage abundance of different species of earthworms in the 10 collection sites during the survey period (1997–1999) is shown in Figures 1 and 2. In most of the study sites, that is 1–7, *L. mauritii* was the dominant species and it showed its presence during the premonsoon, monsoon, and postmonsoon months. *P. corethrurus* showed its abundance in the sites 8–10. Various parameters, that is pH, electrical conductivity (EC), organic carbon (OC), nitrogen (N), atmospheric temperature (AT), soil temperature (ST), soil moisture (SM), humidity (HUM), and rainfall (RF) observed during the survey period (1997–1999) are given in Table 5 and in Figure 3. All the parameters showed fluctuations in all the ten study sites. Here, for the convenience of statistical analysis the parameters were categorized into two major groups: (a) physicochemical parameters that included pH, EC, OC, and N; and (b) climatic parameters which included ST, SM, HUM, and RF.

TABLE 5 Physicochemical and climatic characteristics (average) of the study sites 1 to 10 (1997–1999) (refer to Table 6 for study site description) [22].

Parameter observed*	Study sites									
	1	2	3	4	5	6	7	8	9	10
1997										
pH	7.78	7.63	7.13	6.86	7.59	7.67	6.78	7.04	7.50	6.55

TABLE 5 *(Continued)*

Parameter observed*	Study sites									
	1	2	3	4	5	6	7	8	9	10
EC (dS/m)	0.34	0.20	0.38	0.11	0.21	0.18	0.32	0.14	0.27	0.39
OC (%)	1.42	2.29	4.44	2.75	1.47	2.94	3.42	3.05	4.20	7.99
TN (%)	0.41	0.35	0.23	0.23	0.25	0.40	0.42	0.31	0.22	0.40
ST (°C)	28.90	29.83	27.84	29.29	30.31	29.27	29.83	23.47	22.49	21.30
SM (%)	8.10	6.34	10.34	7.22	7.16	8.30	10.25	15.75	15.46	14.99
1998										
pH	7.95	7.51	7.25	6.66	7.51	7.62	6.45	7.15	7.34	6.44
EC (dS/m)	0.36	0.22	0.38	0.13	0.25	0.16	0.33	0.18	0.31	0.39
OC (%)	1.74	2.19	4.24	2.79	1.43	2.35	4.25	3.19	4.22	8.48
TN (%)	0.44	0.34	0.24	0.27	0.27	0.41	0.41	0.27	0.27	0.38
ST (°C)	29.23	30.18	28.27	30.63	29.92	29.30	30.18	24.14	22.68	21.30
SM (%)	12.14	9.93	15.18	9.73	12.83	12.03	14.35	16.00	17.09	16.29
1999										
pH	7.85	7.49	7.37	6.85	7.38	7.59	6.47	6.98	7.45	6.64
EC (dS/m)	0.35	0.23	0.34	0.14	0.24	0.17	0.35	0.16	0.26	0.39
OC (%)	1.40	2.45	4.34	2.90	1.36	3.02	4.05	3.45	4.37	9.99
TN (%)	0.46	0.37	0.27	0.26	0.26	0.43	0.41	0.31	0.28	0.39
ST (°C)	27.42	28.55	26.49	29.66	30.42	27.46	28.55	25.21	23.51	22.80
SM (%)	9.50	7.27	11.70	8.50	9.27	9.37	10.34	12.56	14.37	13.86

*EC: Electrical conductivity; OC: Organic carbon; TN: total nitrogen; ST: Soil temperature; SM: Soil moisture.

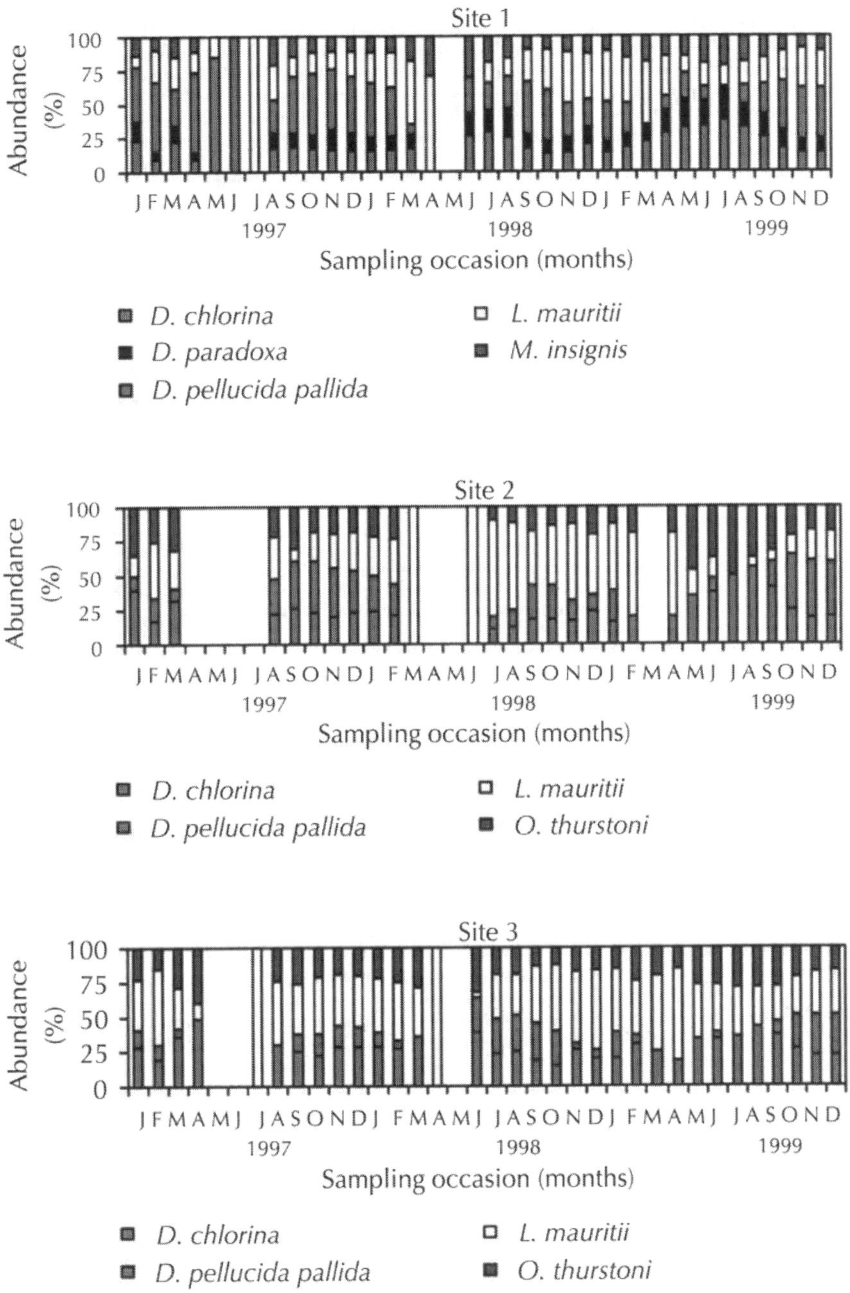

FIGURE 1 *(Continued)*

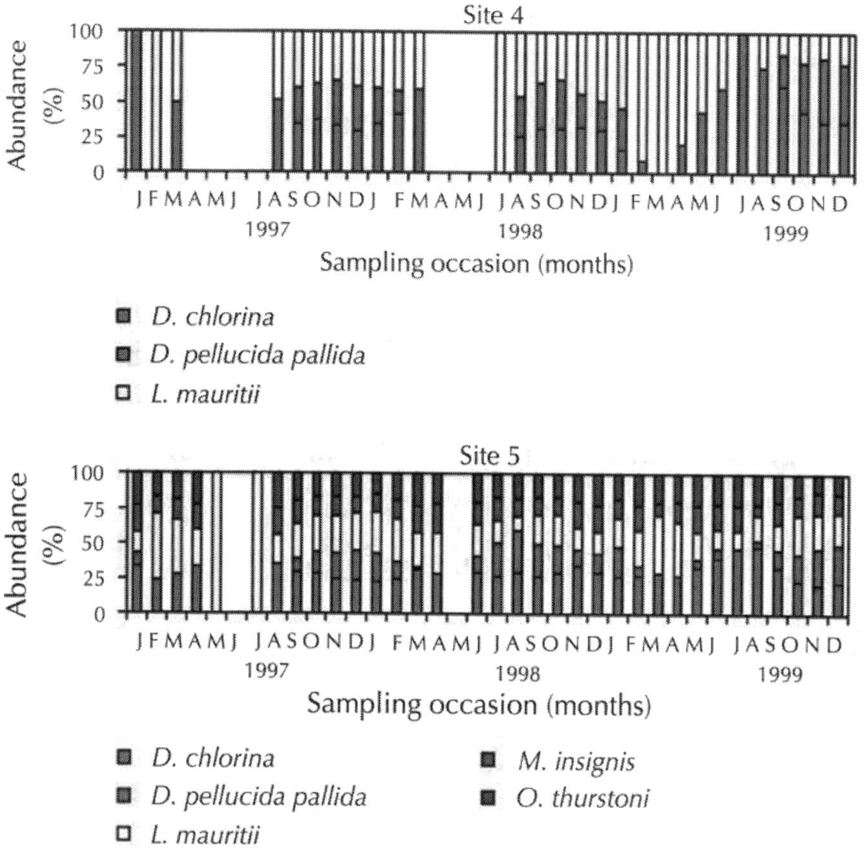

FIGURE 1 Percentage abundance of earthworm population in study sites 1–5 (1997–1999).

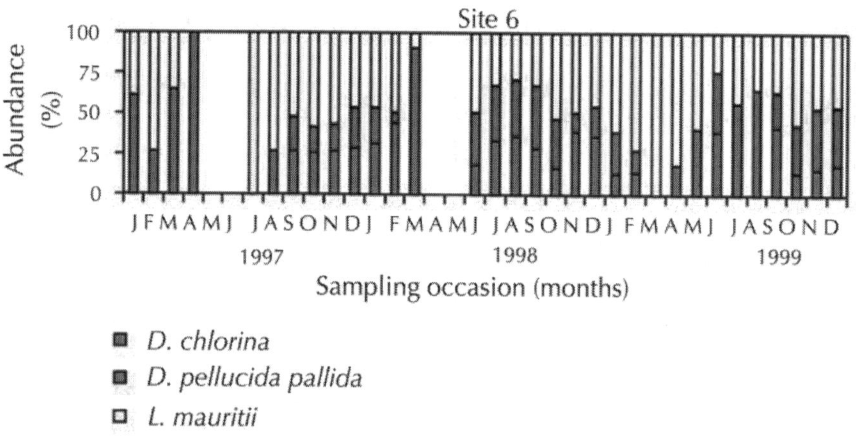

FIGURE 2 *(Continued)*

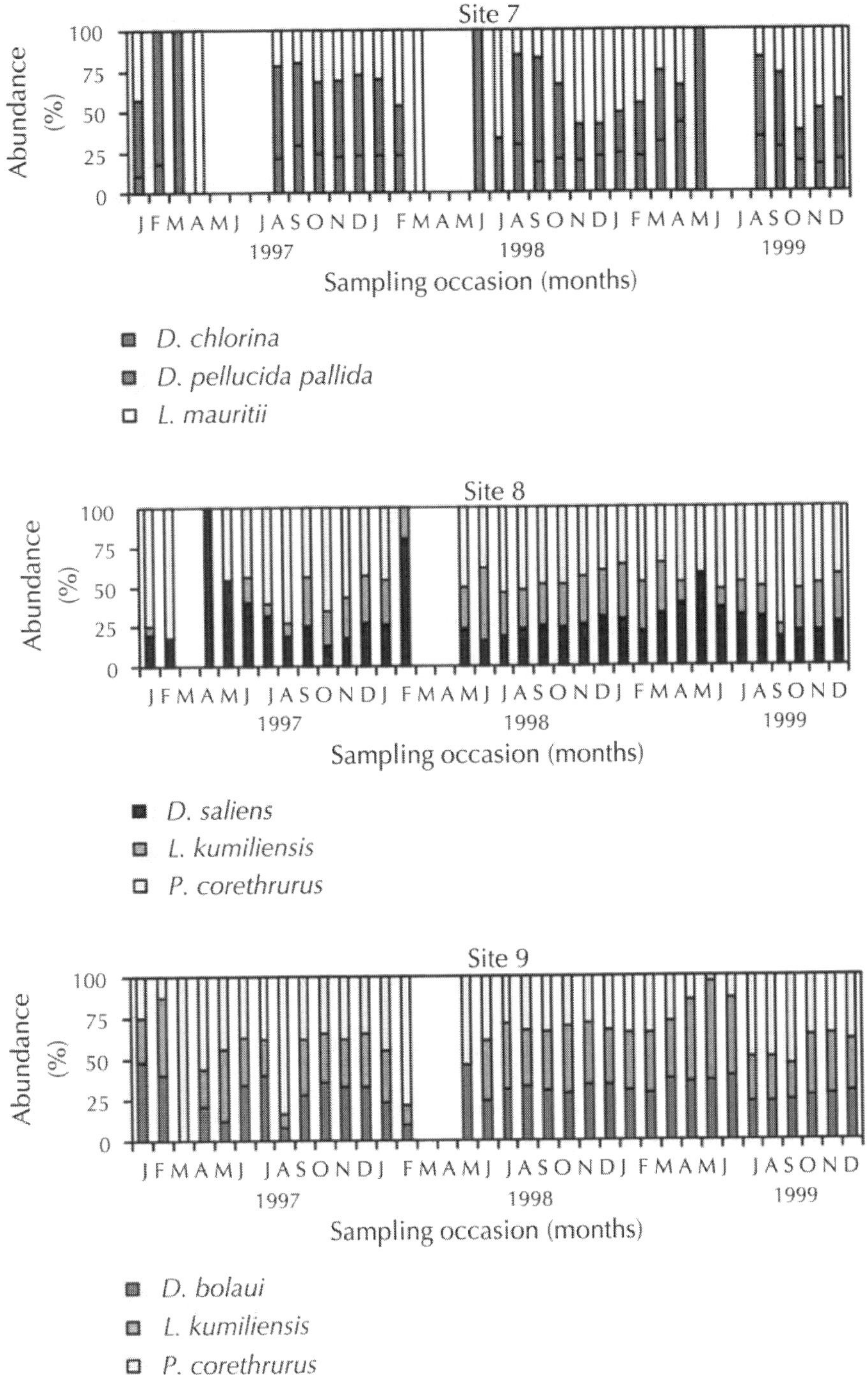

FIGURE 2 *(Continued)*

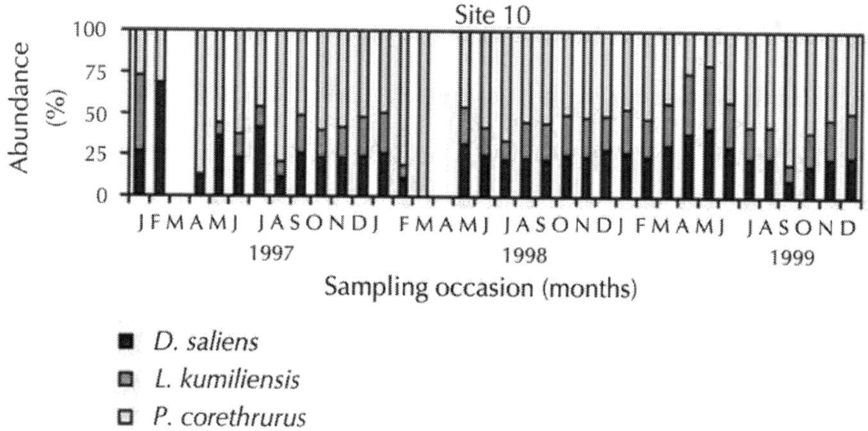

FIGURE 2 Percentage abundance of earthworm population in study sites 6–10 (1997–1999).

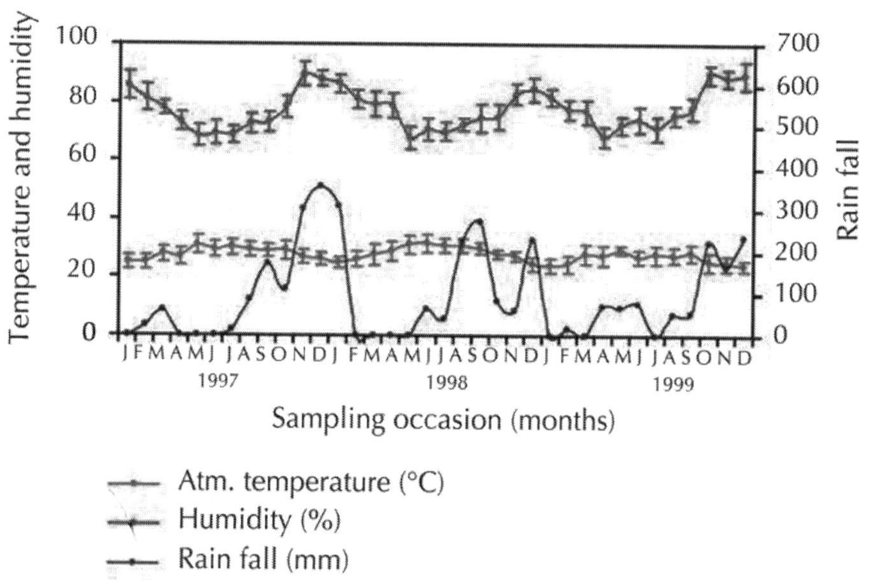

FIGURE 3 *(Continued)*

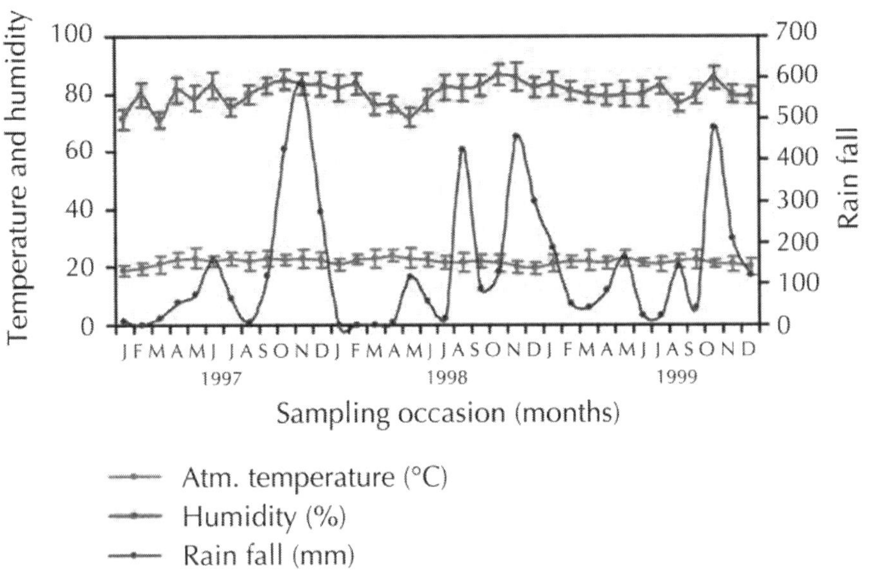

FIGURE 3 The AT (mean ± SD), HUM (mean ± SD), and average RF of the study sites 1–7 (a) and 8–10 (b).

In Tamil Nadu, India, very limited information is available on the distribution pattern of earthworms. The data on earthworm distribution is available for the stations like Palni Hills [24], Madras [25], and Sirumalai Hills [11, 23, 26, 27]. Dindigul, a District in Tamil Nadu, was considered as study site for its variety of habitats to assess the earthworm species diversity, density, and biomass. The population and biomass dynamics of different earthworm species and their percentage abundance in relation to physicochemical characteristics of the soil and the climatic factors were recorded in selected sites. The correlation of earthworm population to physicochemical characteristics of the soil and the climatic parameters was carried out to find out the possibility of arriving at a suitable endemic earthworm species for vermicomposting operations in this part of the country. Since the populations of earthworms are extremely variable in size ranging from only a few individuals (sometimes totally absent) to more than 1000/m^2, the assessment of the size distribution and structure of earthworm population is difficult. The seasonal change, demography, and vertical distribution of the populations make it more complicated, and hence, it is absolutely essential to follow a uniform method of determining the number of earthworms in small sample areas as it has been done in this study. The regular monthly survey carried out for 3 years (1997–1999) showed the presence of ten species of earthworms, with four species restricted only to the hilly region and six species to the plain, including the foothills (Table 6). This observation indicates that species such as *L. kumiliensis, D. bolaui, D. saliens*, and *P. corethrurus* are specific only to the hilly region and they are not found in the foothills. Though *L. kumiliensis* and *L. mauritii* both belong to the same genus, *Lampito, L. kumiliensis* was found only in the hilly region and *L. mauritii* in the plains. This observation indicates that the distribution of different earthworm species is limited even though they are closely related. Such niche differences for closely related species have been reported by earlier workers in the field [28, 29].

TABLE 6 Population density of earthworms in different habitats in Dindigul District, Tamil Nadu studied during 1997–1999 [22].

Study site	Description	*Earthworm species*	Avg. population density (no./m^2)
(1)	Cultivated land	*Lampito mauritii* (Kinb.).	12.52
		Megascolex insignis Mich.	7.82
		Drawida chlorina (Bourne).	8.88
		Drawida paradoxa Rao.	5.10
		Drawida pellucida var. *pallida* Mich.	18.60
(2)	Unirrigated crop land	*Lampito mauritii* (Kinb.).	14.18
		Octochaetona thurstoni Mich.	5.46
		Drawida chlorina (Bourne).	5.04
		Drawida pellucida var. *pallida* Mich.	11.10
(3)	Uncultivated shaded fallow land	*Lampito mauritii* (Kinb.).	17.88
		Drawida pellucida var. *pallida* Mich.	13.27
		Octochaetona thurstoni Mich.	10.92
		Drawida chlorina (Bourne).	10.70
(4)	Uncultivated fallow land	*Lampito mauritii* (Kinb.).	10.30
		Drawida chlorina (Bourne).	4.73
		Drawida pellucida var. *pallida* Mich.	6.46

TABLE 6 *(Continued)*

Study site	Description	*Earthworm species*	Avg. population density (no./m²)
		Lampito mauritii (Kinb.).	15.50
		Megascolex insignis Mich.	10.96
(5)	Garden	*Octochaetona thurstoni* Mich.	13.04
		Drawida chlorina (Bourne).	17.26
		Drawida pellucida var. *pallida* Mich.	11.27
		Lampito mauritii (Kinb.).	8.92
(6)	Orchard	*Drawida chlorina* (Bourne).	6.32
		Drawida pellucida var. *pallida* Mich.	4.75
		Lampito mauritii (Kinb.).	5.63
(7)	Foothills (Alt.< 4 5 0 m.)	*Drawida chlorina* (Bourne).	6.22
		Drawida pellucida var. *pallida* Mich.	22.68
		Lampito kumiliensis (Kinb.).	18.21
(8)	Grassland (Alt. 1,000 m.)	*Dichogaster saliens* (Bedd.).	5.31
		Pontoscolex corethrurus (Muller).	10.30
		Lampito kumiliensis (Kinb.).	29.52
(9)	Semi-evergreen forest (Alt. 1,100 m.)	*Pontoscolex corethrurus* (Muller).	10.49
		Dichogaster bolaui (Mich.).	9.39
		Lampito kumiliensis (Kinb.).	19.42
(10)	Sacred grove land (Alt. 1,300 m.)	*Dichogaster saliens* (Bedd.).	9.37
		Pontoscolex corethrurus (Muller).	19.16

The results of the percentage abundance of different species of earthworms showed that *L. mauritii* and *P. corethrurus* were the most abundant in the study sites 17 and 810, respectively. Formation of aggregation of species has been observed in sites 17, wherever *L. mauritii* was found, it was in association with *D. chlorina* and *D. pellucida pallida*. This sort of association of earthworm species sharing the same habitat is not uncommon [1, 30]. *L. mauritii* is the dominant species found almost all over India along with other earthworm species such as *Drawida modesta, Octochaetona pattoni, O. thurstoni, Ramiella pachpaharensis, Polypheretima elongata*, and *Pontoscolex corethrurus* [8, 31] but Bano and Kale [32] reported that *L. mauritii* was not found in some forest areas and coastal Karnataka. The population densities of earthworms observed in the 10 collection sites ranged from 0 to 228/m^2. Other authors observed population densities (earthworm no./m^2) of 53.5 in plain grass land, 73 in deciduous forest, 543 in the fallow phases of shifting agriculture, and 58.2 in the maize crop land [33-36]. In rubber plantations of Tripura (India) about 20 species of earthworms, namely, *Eutyphoeus gigas, E. gammiei, E. comillahnus, E. assamensis, E. festivas, Eutyphoeus sp., Dichogaster bolaui, D. affinis, Lennogaster chittagongensis, Octochaetona beatrix, Metaphire houlleti, Perionyx sp., Kanchuria sumerianus, Kanchuria sp.1, Kanchuria sp.2, Drawida nepalensis, Drawida sp.1, Drawida sp.2, Pontoscolex corethrurus*, and *Gordiodrilus elegans* were distributed and it was observed that the largely dominating species were endogeics [37].

Evans and Guild [38] have shown that nitrogen rich diets help in rapid growth of earthworms and facilitate more cocoon production than those with little nitrogen available. Due to the influence of nitrogen content of the soil, the percentage contribution of nitrogen to earthworm population might have shown a very high degree of dependence in the present study. Some of the reports from the country well support qualitative dependence of earthworm population on soil nitrogen content [26, 27, 39, 40].

The SM plays a major role in the distribution and occurrence of various earthworm species. The same has been observed by other workers in their studies [25, 28, 29, 41, 42]. The abundance and species diversity are dependent on climatic conditions, especially the occurrence of dry and/or cold periods, and regional variation in vegetation, soil texture, and nutrient content. The climatic parameters, that is, soil temperature, SM, HUM, and RF show seasonal fluctuations (Table 6 and Figure 3). The highest RF was recorded during October-November and the earthworm population was also the highest at this period. The SM content corresponded with earthworm population. Total annual RF of 1130, 1284, and 959 mm was recorded during 1997, 1998, and 1999 in the plains and foothills of Sirumalai (study sites 1–7). The highest RF of 304 and 357 mm was received during October and November 1997 in the study sites. The highest RF months in Sirumalai Hills (study sites 8–10) were October to December. The SM content directly matched with the RF. The SM content ranged from 2.0 to 30.4% in the study sites 7–10 during the 3 years of the study. The HUM also showed fluctuations in both the plains and hilly region of the study area. The SM can explain the increase in earthworm population, since soils are moist under a mulch cover because of the restricted evaporation. There are many indications, to show that the population of endogeic earthworms is controlled mainly by SM [42].

The influence of climatic factors on the populations of earthworm is not uncommon. The populations of *Millsonia anomala* are dependent on climatic conditions as well as vegetational patterns. Earthworm activity and populations are determined essentially by the moisture content of the soil [43]. The temperature and moisture are usually inversely related and higher surface temperature and dry soils are limiting factors to earthworms than low and water-logged soils [44]. The ST plays an important role in the maintenance of earthworm population in an ecosystem and available information also indicates the negative correlation of ST to earthworm population [11, 25, 40, 45]. In rubber plantations of Tripura (India), the earthworms experienced 25.9°C, 24.8%, 4.85, and 1.8% mean soil temperature, moisture, pH, and organic matter, respectively [37]. Temperature largely affects activity of earthworms in temperate regions. Tropical species can withstand higher temperatures. *L. mauritii* is available throughout the year where the annual temperature is 30 ± 2°C. Population of *O. serrata* was active between 27 and 28°C. In tropical regions the temperature fluctuations are minimal when compared to temperate regions.

Moisture is another limiting factor for earthworm distribution as water constitutes a major portion of the body weight of an earthworm. The SM and population estimates are positively correlated [35]. Water constitutes 75–90% of the body weight of earthworms. So the prevention of water loss is a major factor for their survival. They apparently lack a mechanism to maintain constant internal water content, so that their water content is influenced greatly by the water potential of the soil [46], which directly depends on the adequate availability of SM.

The seasonal dynamics in an annual cycle shows that earthworm numbers and biomass were high in the rainy season with a gradual decline in number in the winter season. Earthworms were completely absent during the second half of January and February, when ST was very low (4.9–6.2°C). Dash and Patra [7] and Kale and Krishnamoorthy [8, 47] have recorded maximum number of earthworms and biomass in the rainy and late rainy period. The relationship between earthworm activity and RF was observed by Fragoso and Lavelle [48] and Joshi and Aga [49]. The moisture requirements for different species of earthworms from different regions can be quite different [42]. The dependence of earthworm population on SM is seen in the studies carried out for 3 years as of the highest degree when compared with other climatic parameters. This is because of certain physiological activities of earthworms such as cutaneous respiration and excretion of nitrogenous ammonia and urea, which need a moist environment, which, in turn, is essential for the maintenance of their life process.

Systematic correlation analysis results indicate that only about 80% of the population dependence can be explained by these physicochemical and climatic parameters and it is presumed that the remaining may depend on other environmental factors. The correlation analysis technique may be used to quantify and rationalize the effects of physicochemical parameters on the earthworm population. However, no single factor is likely to be solely responsible for the horizontal distribution of earthworms, but rather the interaction of several of the factors provides suitable soil conditions for the existence of earthworm populations [11].

7.6 EARTHWORM CASTS: ABUNDANCE, STRUCTURE, AND PROPERTIES

Earthworms' release "cast" at the opening of their burrows. Epigeic earthworms release the castings exclusively on soil surface. Their castings may be granular or spindle like masses that may be 23 cm high heaps as in *Eudrilus eugeniae* or *Perionyx excavatus*. There is no definite shape to the excreted matter to identify as castings of *Eisenia fetida*. *Eisenia fetida* releases fine, powdery, dark brown material as surface cast. Soil living endogeic earthworms that feed on different quantities of organic matter along with soil particles use part of their castings to strengthen their burrow walls and the rest is released as castings. Castings of these earthworms may be ovoid or irregularly shaped minute mounds. Though the nature of cast released is characteristic of a species, this cannot be criterion for their identification [50]. If pellet-like castings are released by *Pheretima posthuma, Perionyx millardi* releases threadlike castings. Thick and long winding columns of hollow mound of 5 cm long and 2.5 cm wide casts are characteristic of *Hoplochaetella khandalaensis*. The biggest cast of *Notoscolex birmanicus* weighing 1.6 kg after drying for 4 months is reported from Burma [50]. *Polypheretima elongata* and *Pontoscolex corethrurus* excrete the ingested soil as sticky, thick lumps on soil surface.

Amount of cast produced can serve as an index for assessing earthworm activity. Immediately after rains, release of surface casts will be at a maximum level. At this point of time, majority of earthworms are found at 0 to 10 cm depth and very few of them are found at 20 to 30 cm depth (Kale and Dinesh, 2005, unpublished). Surface cast production has been quantified in different agroecosystems to relate it to their abundance [51-53]. The influence of seasonal variation and land use pattern was observed with respect to cast production in shifting agriculture [34]. Norgrove and Hauser [54] have recorded around 3035 t/ha of cast production in tropical silvicultural system. Reddy [55] has reported annual production of 23.4 to 140.9 tons by *Pheretima alexandri*. According to Lavelle [56], cast production is rhythmic and it will be at maximum at early morning hours. In general cast production in tropical countries is restricted to wet seasons. Table 7 provides the information on earthworm cast production in different agroecosystems during onset of postmonsoon season in the state of Karnataka, India.

TABLE 7 Earthworm cast production during early postmonsoon period (Nov. 2004) at different agroecosystems in Kuti village of Somavarpatna Taluk of Karnataka State (Kale and Dinesh 2004, unpublished).

Land uses	Castings (Kg/Sq. M)
Natural forest	11.20 ± 0.46
Coffee plantation	17.2 ± 0.53
Cardamom plantation	16.80 ± 1.00

TABLE 7 *(Continued)*

Land uses	Castings (Kg/Sq. M)
Paddy fields (after harvest)	13.60 ± 1.00
Acacia plantation	2.40*
Grassland	0.8*

*Due to dryness prevailing at the collection spots castings could be collected only from single spots out of 6 and 8 monolith points.

The physicochemical properties of casts depend on the habitat soil and species of earthworm [57]. Their aggregate stability depends on the available organic matter [58]. The stability of casts and stability of fragmented casts on disintegration are the important factors to determine the soil structure [1]. Aggregate stability may result from addition of mucus secretion from earthworm gut and of associated microorganisms in the gut. It may also be due to macerated organic particles in the castings that encourage microbial activity after its release from the gut [59]. According to Parle [60], stability of casts is due to fungal succession that takes place in the cast. Habibulla and Ismail [61] are of the opinion that soil texture, particle size, and porosity play an important role in burrowing and surface cast production. As casting activity is restricted to wet seasons, not much of attention is paid to assess the quantum of cast produced and its influence on soil physical, chemical, and biological properties as is available from other parts of the world. It is essential to know the physicochemical and biological variations that may be seen in cast produced by the same species of earthworm inhabiting places that differ in physiographic and edaphic conditions. This will provide the information on interrelationship of earthworms, original soil characters, and nature of available organic material that influence the change in soil characters through deposition of earthworm cast. The fertile lands turning unproductive in Himachal Pradesh, India, due to sticky castings of earthworms that turned the soil into cement-like clods had been reported [62]. Puttarudraiah and Sastry [63] had observed stunting of growth in root crops like carrot, radish, and beetroots due to castings of Pontoscolex corethrurus in pot culture studies.

Castings of earthworms are the "store house" of nutrients for plants. The increased earthworm activity, with increase in availability of carbon and in turn a raise in available nitrogen and phosphorus in their castings was also reported [6]. Earthworm activity has shown to improve the soil aggregates and soil minerals that are more available to plants than from soil [54, 64]. It is clear from various studies that earthworm casts may have more important role in plant nutrition and nutrient cycling than it was assumed [65, 66]. In India, very early reports are available on such observations on the chemical properties of earthworm castings that can play a positive role in plant growth [57, 67, 68]. The chemical composition of casts, which is widely studied, is of holonephric lumbricid earthworms. In subtropical country like India where majority of earthworms are meronephric, their castings may show higher level of available plant nutrients than surrounding soil. Dash and Patra [7, 53] had reported higher levels of nitrogen in casts of *Lampito mauritii* than in surrounding soil. Ganeshmurthy et al. [69] have found higher rate of mobilization of micronutrients in earthworm castings. It requires further studies on meronephric Megascolecid earthworms and their castings on available and exchangeable forms of nutrients to assess their contribution to soil fertility. Kale and Krishnamoorthy [70] had shown increased levels of soluble calcium and carbonates in castings of *Pontoscolex corethrurus*. Soluble carbonates contribute to exchangeable base contents of castings (Table 8). The physicochemical properties like pH, EC, organic C, total N, available P, K, Na, Ca, and Mg of casts did not differ in zero tillage land treated with mulch of residues of annuals or perennials [19]. The population dynamics of a peregrine earthworm, *Pontoscolex corethrurus*, in undisturbed soil of Sirumalai Hills clearly showed that the parameters like RF, HUM, SM, and OC influence the population positively [26, 27]. It has also been reported that in rubber plantations of Tripura, a part of North-East India, *Pontoscolex corethrurus* was the dominant species, representing 61.5% biomass and 72% density of the total earthworm population where it might be linked to individual tree species effect (*Hevea brasiliensis*) that favored *P. corethrurus* over other species [37].

TABLE 8 Calcium and carbonates in castings of *Pontoscolex corethrurus* compared with that of habitat soil [70].

Constituents	µg/g dry weight	
	Soil	Castings
Ionic Calcium	12.24 ± 0.41	145.50 ± 9.81
Exchangeable Calcium	12.83 ± 0.37	95.23 ± 7.28
Insoluble Calcium	179.62 ± 0.02	32.09 ± 0.93
Ionic/Insoluble Carbonate	0.15 ± 0.01	6.98 ± 2.22

7.7 EARTHWORMS AND MICROFLORA

Earthworm activity is closely associated with microbial activity. Lavelle [2] is of the opinion that there may exist competition between microorganisms and earthworms for easily digestible and energy rich substrates. Such competition may depend on availability of nutrients in the medium. Contrary to this, earthworms may derive benefit from microorganisms when they have to survive on materials rich in cellulose or hemi cellulose. So there exists mutualistic relation between earthworms and microorganisms. Tiunov and Scheu [6] have shown that earthworms deprive easily available carbon to microorganisms and availability of carbon increases effective mobilization of N and P by earthworms. The complex interrelationship of earthworms and microorganisms is at the level of their digestive tract, castings, and burrow walls [71]. This establishes the probable mutualism that exists between earthworms and microorganisms. Joshi and Kelkar [68] demonstrated higher microbial activity in earthworm castings and their role in mineralization of nitrogen. They incubated known weights of groundnut cake in a pot containing earthworm castings and other containing soil from the same place. The release of N from groundnut cake was at a higher level in pot containing castings than from one having soil as the medium.

Bhat, et al. [72] were the pioneer contributors to report on role of microorganisms in the gut of earthworms. Khambata and Bhat [73] had made a detailed investigation on intestinal microflora of *Pheretima sp*. They had isolated Pseudomonas, Corenyform bacteria, Nocardia, Streptomyces, and Bacillus from the intestinal tract. There is no report of nitrosofying and nitrifying bacteria in their observations in the gut of earthworms. Dash et al. [74] have reported about isolation of 16 fungi from different parts of the gut out of 19 found in their habitat. In the fresh castings of the same earthworms there were only seven fungi with antibiotic properties or with thick spore coats. This suggests the selective fungal feeding by earthworms.

Drillosphere is the focus for understanding earthworm microbe interrelationship. This association is also associated with land use and metabolizable carbon present in the soil. Metabolizable carbon has positive effect on both microorganisms and earthworms [75]. Microbial activity will be at a higher level in the drillosphere than in surrounding soil and other edaphic factors determine the microbial diversity in drillosphere [76]. According to Kretzschmar [77], interaction of soil fauna and microflora determines soil dynamics. The contribution of their activity for formation of humus is an index for soil fertility. Bhatnagar [78] had expressed that at 2040 cm depth in drillosphere zone there were 40% aerobic N-fixers, 13% anaerobic N-fixers, and 16% of denitrifiers. He attributed low C/N ratio in soils rich in earthworm population because of stimulation of N-fixers in drillosphere. Drillosphere provides necessary substrate for growth and establishment of microorganisms.

Recent developments in the country as well as at the global level are the application of detritivorous epigeic earthworms for organic manure/vermicompost production from biodegradable organic materials recovered from agricultural lands, agro-based industries, and municipal solid waste. This field of study is closely associated with earthworm microbe interaction. The quality of the manure or vermicompost depends on microorganisms associated with the process of decomposition. Bhat [79] had reported that the diet formulation or the composition of organic matter used as feed influences the microflora associated with earthworm activity. Similar studies were made on enhanced N-fixers activity on using 2% neem cake in the feed mix of earthworm *Eudrilus eugeniae* [80] (Table 9).

TABLE 9 Microbial population in neem cake enriched vermicompost [80].

Microbial population no./g vermicompost	Vermicompost with 2% neem cake	Vermicompost without neem cake
Fungi no. × 10^4	22.3	5.2
Bacteria no. × 10^6	15.0	7.8
Nitrogen fixers no. × 10^5	54.1	6.6

During winter months in Himalayan region, fungal population was higher in vermicomposting system than in the native soil [81]. The maintenance of temperature in vermicomposting system at a favorable level for earthworm activity might have been the reason for establishment of fungal population. Press mud, a byproduct of the sugar industry, is often used as one of the substrates in vermicomposting. Subjecting of this material to earthworm activity along with other organic matter has resulted in changes in microbial populations [82]. Rajani et al. [83] have related the microbial density and enzyme activity as a measure to assess the effectiveness of process of vermicomposting. It is essential to make an in-depth study to understand the mutualistic association between microflora and earthworms in mechanism of decomposition of organic matter. An increase in actinomycetes population was observed in the gut region of earthworms. Some of the isolates from gut region of earthworms have expressed growth stimulatory effect when used in pot cultures of tomato and finger millet [84].

The colony forming units (CFUs) of bacteria and fungi in the casts of *P. corethrurus* significantly deviated from the CFU found in adjacent soil. The correlation between the physicochemical parameters and microbial populations of the casts of *P. corethrurus* showed that the establishment of microbial population requires optimum moisture, OC, and nitrogen content [20]. The vermicasts of *P. ceylanensis* showed 14 different fungal species belonging to the genera, Aspergillus, Chaetomium, Cladosporium, Cunninghamella, Fusarium, Mucor, Penicillium, and Rhizopus. The TN, P, K, Ca, Cu, Fe, and Zn were higher in vermicasts than in control (substrate without earthworms) while OC and C/N ratio were lower in vermicasts. The total OC was 42.3% in the control whereas it was 35.2% in the vermicasts of *P. ceylanensis*. The incubation of vermicasts (45 days) showed significant correlation with that of the increase in fungal population ($r = 0.720$; $P < 0.05$) and decrease in moisture content ($r = -0.984$; $P < 0.001$), and the decrease in moisture content statistically had no effect on the total fungal population in the vermicasts of *P. ceylanensis* [85]. The total microbial population, namely, bacteria, fungi, and actinomycetes was found to be manyfold higher than in the initial vermibed substrate and in substrate without earthworms (control). The initial count of bacteria, fungi, and actinomycetes in the control was 123.42 CFU $\times 10^7 g^{-1}$, 159.64 CFU $\times 10^3 g^{-1}$, and 86.90 CFU $\times 10^4 g^{-1}$ whereas in castings (vermicompost) of *P. ceylanensis* the reported microbial populations were 268.62, 223.39, and 141.09 [86]. These observations clearly indicate the importance of microorganisms associated with earthworms in creating suitable environment for the standing crops as well as for vermicomposting of different organic wastes. It is still at the infancy to draw any inference regarding earthworm, microbe, and plant association.

Studies are also in progress to assess the inhibitory effects of the principles present in the body wall, gut extract, and of coelomic fluid on some selected plant and animal pathogens.

7.8 EARTHWORMS AS BIOINDICATORS

Earthworms can also serve as indicators of several changes/factors associated with soil. Many studies clearly showed that the earthworms are best indicators of heavy metals, toxic pollutants, and direct and indirect anthropogenic changes in soil [87-89]. A study conducted in northern semiarid region of India showed the presence of earthworms to the maximum level wherever the farmers followed integrated farming (100%) practice and this was followed by organically managed (70%) and conventional (18.9%) agroecosystems. The earthworm abundance was directly related to the management practices and the values of ecological indices like Shannon diversity (H'), species dominance (C), the species richness (S), and evenness (E). This clearly illustrates the anthropogenic pressure on earthworm communities in arable lands [90]. A similar report from Ivory Coast is available on the impact of land use changes and land use intensification on earthworm populations and diversity in intermediate disturbed systems [91]. Even though these studies suggest the use of earthworms as bioindicators of man-made changes, it necessitates more field and laboratory investigations to find out earthworm community structure, species interrelations, and the most efficient species to be used in biomonitoring of ecosystem degradation due to anthropogenic activities in the forest areas.

Certain toxic substances in soil affect the behavior and physiology of earthworms that can serve as biomonitoring tool for their systematic effect on soil organisms and other higher organisms. For example, the presence of tetra ethyl lead (TEL) in leaded gasoline and lead oxide has a significant effect on behavior, morphology, and histopathology of earthworms. The absorption of TEL into the tissues of earthworms produced severe effects, rupture of the cuticle, extrusion of coelomic fluid, and inflexible metameric segmentation. This led to desensitization of the posterior region and its fragmentation [92].

The efficient potential of earthworms in bioaccumulation of heavy metals in their tissues serves as ecological indicator of soil contaminants. As per the recent report from India, the level of DTPA extractable metals in casts of earthworms, *Metaphire posthuma* (endogeic) and *Lampito mauritii* (anecic) collected from cultivated land, urban garden and sewage soils were higher than those of surrounding soil. The concentration of Zn, Fe, Pb, and Mn in earthworm casts was higher in sewage soil followed by cultivated land and urban garden, respectively. There exists a close relationship between metal concentration in earthworm tissues and surrounding soils. The study also revealed the presence of species specificity in metal accumulation in earthworms. Higher level of metal concentrating in the tissues was found in endogeic *M. posthuma* than in tissues of anecic *L. mauritii*. The difference in burrowing patterns may influence the patterns of bioaccumulations of metals apart from other contributory factors. Further, more detailed study is still required to elaborate the proposed hypothesis [93]. Analogous study conducted in Egypt also suggests that the variation in heavy metal concentration in soil and earthworms in different sites may be significant depending on soil properties and pollution status [88]. Sizmur and Hodson [94] evidently suggested that earthworms increase metal mobility and availability but more studies are required to determine the precise mechanism for this. So, this field of research with earthworm requires in depth research to understand the functional role of earthworms as bioindicators and bioconcentrators.

7.9 EARTHWORMS AND VERMICOMPOSTING: INDIAN SCENARIO

The familiar earthworm species, *Eudrilus eugeniae, Eisenia fetida, Lumbricus rubellus*, and *Perionyx excavatus*, are well known for their efficiency in vermicomposting.

It is desirable to know about other species of earthworms that may be as efficient or better in their performance over the mentioned species in a country having rich diversity of fauna for *in situ* and *ex situ* vermiculture. There are more than a dozen of earthworm species that have been reported to be efficient in vermicomposting. Most of the species that are included under genus Perionyx show great potential to work on organic matter. Apart from the well-known P. excavatus, other Perionyx species such as *P. ceylanensis*, *P. bainii, P. nainianus*, and *P. sansibaricus* are recently considered to be the potential vermicomposting earthworms [20, 95-97]. Future investigations provide scope for identifying more species with vermicomposting potential.

In natural systems, if earthworms are ecosystem engineers, in manmade seminatural systems of organic residues, the detritivorous earthworms are saviors of biosphere from organic pollutants. From the review, it is very clear that the earthworm ecology needs much attention with reference to their functional role in different ecosystems. By the way of exploration, it might be possible to understand the significant role of earthworms in plant-microbe interactions. With regard to vermiculture, it is necessary to work on the idea of developing the consortia of earthworm species for vermiculture practices in India. It is always better to develop and encourage polycultures rather than maintaining monoculture. Moreover, with diversity in agricultural residues and by-products from agro industries, it is essential to identify earthworms that will accept these materials with minimum effort and investment.

7.10 CONCLUSION

There are more than 500 species of earthworms distributed in different geographical regions in India, in different ecosystems. Being partly subtropical and partly temperate, majority of earthworms are endogeic or geophagous. Even among the epigeic earthworms (ca. 8%), those that are voracious feeders, are efficient biomass producers, and have short life cycle, high rate of fecundity, and high rate of adaptability to changing physicochemical properties of feed material can only serve as successful species for vermiculture. One has to look for these characters before recommending any species for vermiculture. The species that is promising under protected laboratory conditions in a small scale may fail to perform under field conditions when it is expected to work on large amount of organic matter. The present scenario in India shows that there is good response from the farmers to adopt the technology for producing vermicompost to use as soil amendment. They are reaping the benefits of using the recommended species for producing required quantity of vermicompost to fulfill the needs of their land and also to market the production to other neighborhood farmers. Still many avenues remain open for the scientists to carry out research in this field to unravel various problems associated with the technology.

The studies are at preliminary stages and it will require some more time to draw any conclusions based on the available data. Such interdisciplinary applications of earthworm research help to understand the functional complexity of these organisms other than their contribution to management of organic biodegradable residues as the major secondary detritivorous group.

KEYWORDS

- **Agroecosystems**
- **Detritivorous**
- **Macrofauna**
- **Microorganisms**
- **Vermicomposting**

ACKNOWLEDGMENT

The study was financed by the Natural Sciences and Engineering Research Council of Canada *via* a Strategic Grant to Dr. Joann Whalen, Dr. Andrew Gordon and others. The authors would like to thank Dr. Joann Whalen from McGill University and Dr. Rick Bourbonniere from Environment Canada for all of their technical assistance.

REFERENCES

1. Lee, K. E. *Earthworms: Their Ecology and Relationships with Soils and Land Use.* Australia Academic Press, Sydney (1985).
2. Lavelle, P. The structure of earthworm communities. In: *Earthworm Ecology*, J. E. Satchell (Ed.). Chapman and Hall, London UK, pp. 449–466 (1983).
3. Satchell, J. E. and Lowe, D. G. Selection of leaf litter in *Lumbricus terrestris*. In: *Progress in Soil Biology*, O. Graff and J. E. Satchell (Eds.). North Holland, Amsterdam Netherlands, pp. 102–119 (1967).
4. Ganihar, S. R. Nutrient mineralization and leaf litter preference by the earthworm *Pontoscolex corethrurus* on iron ore mine wastes. *Restoration Ecology*, **11**(4), 475–482 (2003).
5. Kale, R. D. and Krishnamoorthy, R. V. Litter preferences in the earthworm *Lampito mauritii. Proceedings of Indian Academy Sciences*, **90**, 123–128 (1981).
6. Tiunov, A. V. and Scheu, S. Carbon availability controls the growth of detritivores (*Lumbricidae*) and their effect on nitrogen mineralization. *Oecologia*, **138**(1), 83–90 (2004).
7. Dash, M. C. and Patra, V. C. Density, biomass, and energy budget of a tropical earthworm population from a grassland site in Orissa, India. *Revue d'Ecologie Et de Biologie du Sol*, **14**, 461–471 (1977).
8. Kale, R. D. and Krishnamoorthy, R. V. Cyclic fluctuations and distribution of three species of tropical earthworms in a farmyard garden in Bangalore. *Rev d' Ecolgie et de Biol du Sol*, **19**, 67–71 (1982).
9. Reddy, M. V., Kumar, V. P. K., Reddy, V. R., et al. Earthworm biomass response to soil management in semi-arid tropical Alfisol agroecosystems. *Biology and Fertility of Soils*, **19**(4), 317–321 (1995).
10. Tripathi, G. and Bhardwaj, P. Seasonal changes in population of some selected earthworm species and soil nutrients in cultivated agroecosystem. *Journal of Environmental Biology*, **25**(2), 221–226 (2004).
11. Karmegam, N. and Daniel, T. Effect of physico-chemical parameters on earthworm abundance: a quantitative approach. *Journal of Applied Sciences Research*, **3**, 1369–1376 (2007).
12. Sathianarayanan, A. and Khan, A. B. Diversity, distribution, and abundance of earthworms in Pondicherry region. *Tropical Ecology*, **47**(1), 39–144 (2006).
13. Kale, R. D., Bano, K., and Krishnamoorthy, R. V. Anachoresis of earthworms. *Bombay Natural History Society*, **78**, 400–402 (1981).

14. Subler, S., Baranski, C. M., and Edwards, C. A. Earthworm additions increased short-term nitrogen availability and leaching in two grain-crop agroecosystems. *Soil Biology and Biochemistry*, **29**(3–4), 413–421 (1997).
15. Robinson, C. H., Ineson, P., Piearce, T. G., and Rowland, A. P. Nitrogen mobilization by earthworms in limed peat soils under Picea sitchensis. *Journal of Applied* Ecology, **29**(1), 226–237.
16. Edwards, C. A. and Lofty, J. R. Nitrogenous fertilizers and earthworm populations in agricultural soils. *Soil Biology and Biochemistry*, **14**(5), 515–521 (1982).
17. Knight, D. P., Elliott, P. W., Anderson, J. M., and Scholefield, D. The role of earthworms in managed, permanent pastures in Devon, England. *Soil Biology and Biochemistry*, **24**(12), 1511–1517 (1992).
18. Shipitalo, M. J. and Butt, K. R. Occupancy and geometrical properties of *Lumbricus terrestris* L. burrows affecting infiltration. *Pedobiologia*, **43**(6), 782–794 (1999).
19. Reddy, M. V., Reddy, V. R., Balashouri, P., et al. Responses of earthworm abundance and production of surface casts and their physico-chemical properties to soil management in relation to those of an undisturbed area on a semi-arid tropical alfisol. *Soil Biology and Biochemistry*, **29**(34), 617–620 (1997).
20. Karmegam, N. and Daniel, T. Selected physico-chemical characteristics and microbial populations of the casts of the earthworm, Pontoscolex corethrurus (Muller) and surrounding soil in an undisturbed forest floor in Sirumalai Hills, South India. *Asian Journal of Microbiology, Biotechnology and Environmental Sciences*, **2**, 231–234 (2000).
21. Julka, J. M., Paliwal, R., and Kathireswari, P. *Biodiversity of Indian earthworms—an overview.* In: Proceedings of Indo-US Workshop on Vermitechnology in Human Welfare. C. A Edwards, R. Jayaraaj, and I. A. Jayraaj (Eds.). Rohini Achagam, Coimbatore, India, pp. 36–56 (2009).
22. Karmegam, N. Studies on earthworms, vermiculture, vermicomposting and utilization of vermicompost for plant growth. Ph.D. thesis. Gandhigram Rural University, Gandhigram, Tamil Nadu, India (2002).
23. Karmegam, N. and Daniel, T. A first report on the occurrence of a Megascolecid earthworm, Lampito kumiliensis (Annelida: Oligochaeta) in Sirumalai Hills of Tamil Nadu, South India. *Ecology, Environment and Conservation*, **7**, 115–116 (2001).
24. Jamieson, B. G. M. Preliminary descriptions of Indian earthworms (Megascolecidae: Oligocaheta) from the Palni hills. *Bullettin du Muséum National d'Histoire Naturelle Section A*, **313**, 478–502 (1977).
25. Ismail, S. A. and Murthy, V. A. Distribution of earthworms in Madras. *Proceedings of Indian Academy of Sciences (Animal Science)*, **94**, 557–566 (1985).
26. Karmegam, N. and Daniel, T. Abundance and population density of three species of earthworms (Annelida: Oligochaeta) in foothills of Sirumalai (Eastern Ghats), South India. *Indian Journal of Environment and Ecoplanning*, **3**, 461–466 (2000).
27. Karmegam, N. and Daniel, T. Population dynamics of a peregrine earthworm, Pontoscolex corethrurus in an undisturbed soil in Sirumalai Hills of Tamil Nadu, South India. *Journal of Ecological Research and Bioconservation*, **1**, 9–14(2000).
28. Bhadauria, T. and Ramakrishnan, P. S. Population dynamics of earthworms and their activity in forest ecosystems of north-east India. *Journal of Tropical Ecology*, **7**(3), 305–318 (1991).
29. Chaudhuri, P. S. and Bhattacharjee, G. Earthworm resources of Tripura. *Proceedings of National Academy of Sciences, India*, **69**, 159–170 (1999).
30. Edwards, C. A. and Bohlen, P. J. *Biology and Ecology of Earthworms*. Chapman and Hall London, UK (1996).
31. Ismail, S. A. Earthworm resources of Madras. In: Proceedings of National Seminar on Organic Waste Utilization and Vermicomposting, *Part B: Verms and Vermicomposting*, M. C. Dash, B. K. Senapati, and P.C. Mishra (Eds.), pp. 8–15 (1986).
32. Bano, K. and Kale, R. D. Earthworm fauna of Southern Karnataka, India. In: Advances in Management and Conservation of Soil Fauna. G. K. Veeresh, D. Rajagopal, C. A. Virakthamath (Eds.). Oxford & IBH, New Delhi India, pp. 627–634 (1991).

33. Krishnamoorthy, R. V. Competition and coexistence in a tropical earthworm community in a farm garden near Bangalore. *Journal of Soil Biology and Ecology*, **5**, 33–47 (1985).
34. Bhadauria, T. and Ramakrishnan, P. S. Earthworm population dynamics and contribution to nutrient cycling during cropping and fallow phases of shifting agriculture (jhum) in north-east India. *Journal of Applied Ecology*, **26**(2), 505–520 (1989).
35. Ismail, S. A., Ramakrishnan, C., and Anzar, M. M. Density and diversity in relation to the distribution of earthworms in Madras. *Proceedings of Indian Academy of Sciences: Animal Sciences*, **99**(1), 73–78 (1990).
36. Kaushal, B. R., Bisht, S. P. S., and Kalia, S. Population dynamics of the earthworm *Amynthas alexandri* (Megascolecidae: Annelida) in cultivated soils of the Kumaun Himalayas. *Applied Soil Ecology*, **2**(2), 125–130 (1995).
37. Chaudhuri, P. S., Nath, S., and Paliwal, R. Earthworm population of rubber plantations (Hevea brasiliensis) in Tripura, India. *Tropical Ecology*, **49**(2), 225–234 (2008).
38. Evans, A. V. and Guild, W. J. Mc. L. Studies on the relationships between earthworms and soil fertility. IV. On the life cycles of some British Lumbricidae. *Annals of Applied Biology*, **35**, 471–484 (1948).
39. Senapati, B. K. and Sahu, S. K. Population, biomass and secondary production in earthworms. In *Earthworm Resources and Vermiculture*, Zoological Survey of India, Calcutta India, pp. 57–78 (1993).
40. Ganihar, S. R. Earthworm distribution with special reference to physicochemical parameters. *Proceedings of Indian National Science Academy B*, **62**, 11–18 (1996).
41. González, G., Zou, X., and Borges, S. Earthworm abundance and species composition in abandoned tropical croplands: comparisons of tree plantations and secondary forests. *Pedobiologia*, **40**(5), 385–391 (1996).
42. Auerswald, K., Weigand, S., Kainz, M., and Philipp, C. Influence of soil properties on the population and activity of geophagous earthworms after 5 years of bare fallow. *Biology and Fertility of Soils*, **23**(4), 382–387 (1996).
43. Lavelle, P. Earthworm activities and the soil system. *Biology and Fertility of Soils*, **6**(3), 237–251(1988).
44. Nordström, S. and Rundgren, S. Environmental factors and Lumbricid associations in Southern Sweeden. *Pedobiologia*, **13**, 301–326 (1974).
45. Senapati, B. K. and Dash, M. C. Influence of soil temperature and moisture on the reproductive activity of tropical earthworms of Orissa. *Journal of Soil Biology and Ecology*, **4**, 13–21 (1984).
46. Kretzschmar, A. and Bruchou, C. Weight response to the soil water potential of the earthworm *Aporrectodea longa*. *Biology and Fertility of Soils*, **12**(3), 209–212 (1991).
47. Kale, R. D. and Krishnamoorthy, R. V. Distribution and abundance of earthworms in Bangalore. *Proceedings of Indian Academy of Sciences*, **88**, 23–25 (1978).
48. Fragoso, C. and Lavelle, P. The earthworm community of a Mexican tropical rain forest (Chajul, Chiapas). In: On Earthworms, A. M. Bonvincini Paglai and P. Omodeo (Eds). *Selected Symposia and Monographs U.Z.I.*, Modena, Mucchi, Italy, pp. 281–295 (1987).
49. Joshi, N. and Aga, S. Diversity and distribution of earthworms in a subtropical forest ecosystem in Uttarakhand, India. *The Natural History Journal of Chulalongkorn University*, **9**, 21–25 (2009).
50. Tembe, V. B. and Dubash, P. J. The earthworms: a review. *Journal of Bombay Natural History Society*, **58**, 171–201 (1961).
51. Roy, S. K. Studies on the activities of earthworms. *Proceedings of Zoological Society*, Calcutta, **10**, 81–98 (1957).
52. Gates, G. E. Ecology of some earthworms with special reference to seasonal activity. *American Midland Naturalist*, **66**, 61–86 (1961).
53. Dash, M. C. and Patra, V. C. Worm cast production and nitrogen contribution to soil by a tropical earthworm population from a grassland site from Orissa, India. *Revue d Ecologie Et de Biologie du Sol*, **16**, 79–83 (1979).

54. Norgrove, L. and Hauser, S. Effect of earthworm surface casts upon aize growth. Pedobiologia, **43**(6), 720–723 (1999).
55. Reddy, M. V. Annual cast production by the Megascolecid earthworm, Pheretima alexandri (Beddard). *Comparative Physiology and Ecology*, **8**, 84–86 (1983).
56. Lavelle, P. Les vers de terre de la savane de Lamto. Analyse d'un Ecosysteme tropical humide: LA Savanede Lamto (Côte d'Ivoire). *Bullettin De Liaison des chercheurs de Lamto*, **5**, 133–136 (1974).
57. Nijhawan, S. D. and Kanwar, J. S. Physicochemical properties of earthworm castings and their effect on the productivity of the soil. *Indian Journal of Agricultural Sciences*, **22**, 357–373 (1952).
58. Dutt, A. K. Earthworms and soil aggregation. *Journal of American Society for Agronomy*, **48**, 407 (1948).
59. Swaby, R. J. The influence of earthworm on soil aggregation. *Journal of Soil Science*, **1**, 195–197 (1950).
60. Parle, J. N. A microbiological study of earthworm casts. *Journal of General Microbiology*, **31**, 13–22 (1963).
61. Habibullah, A. M. and Ismail, S. A. Preference to soil fractions and the effect of soil compaction on the casting and burrowing behavior of the earthworm *Lampito mauritii*. *Journal of Soil Biology and Ecology*, **5**, 26–32 (1985).
62. Agarwal, G. W., Rao, K. S. K., and Negi, L. S. Influence of certain species of earthworms on the structure of some hill soils. *Current Science*, **27**, 213 (1958).
63. Puttarudraiah, M. and Sastry, K. S. S. A preliminary study of earthworm damage to crop growth. *Mysore Agriculture Journal*, **36**, 2–11 (1961).
64. Fraser, P. M., Beare, M. H., Butler, R. C., Harrison-Kirk, T., Piercy, J. E. Interactions between earthworms (*Aporrectodea caliginosa*), plants and crop residues for restoring properties of a degraded arable soil. *Pedobiologia*, **47**(5–6), 870–876 (2003).
65. Callaham, Jr. M. A. and Hendrix, P. F. Impact of earthworms (Diplocardia: *Megascolecidae*) on cycling and uptake of nitrogen in coastal plain forest soils from northwest Florida, USA. *Applied Soil Ecology*, **9**(1–3), 233–239 (1998).
66. Materechera, S. A., Mandiringana, O. T., and Nyamapfene, K. Production and physico-chemical properties of surface casts from microchaetid earthworms in central Eastern Cape. *South African Journal of Plant and Soil*, **15**(4), 151–157 (1998).
67. Shrikande, J. G. and Pathak, A. N. Earthworms and insects in relation to soil fertility. *Current Science*, **17**, 327–328 (1948).
68. Joshi, N. V. and Kelkar, B. V. Role of earthworms in soil fertility. *Indian Journal of Agricultural Science*, **22**, 189–196 (1952).
69. Ganeshamurthy, A. N., Manjaiah, K. M., and Rao, A. S. Mobilization of nutrients in tropical soils through worm casting: availability of micronutrients. *Soil Biology and Biochemistry*, **30**(13), 1839–1840 (1998).
70. Kale, R. D. and Krishnamoorthy, R. V. The calcium content of body tissues and castings of earthworm Pontoscolex corethrurus (*Annelida-Oligochaeta*). *Pedobiologia*, **20**, 309–315 (1980).
71. Edwards, C. A. and Arancon, N. Q. Interactions among organic matter, earthworms and microorganisms in promoting plant growth. In: *Soil Organic Matter in Sustainable Agriculture*, pp. 327–376 (2004).
72. Bhat, J. V., Khambata, S. R., Maya, G. B., Sastry, C. A., Iyer, R. V., and Iyer, V. Effect of earthworms on the microflora of the soil. *Indian Journal of Agricultural Science*, **30**(2), 106–114 (1960).
73. Khambata, S. R. and Bhat, J. V. A contribution to the study of the intestinal microflora of Indian earthworms. *Archives of Microbiology*, **28**(1), 69–80 (1957).
74. Dash, M. C., Mishra, P. C., and Behara, N. Fungal feeding by a tropical earthworm. *Tropical Ecology*, **20**, 9–12 (1979).

75. Dlamini, T. C., Haynes, R. J., and Van Antwerpen, R. Exotic earthworm species dominant in soils on sugarcane estates in the Eshowe area of the north coast of Kwazulu-Natal. In: *Proceedings of Annual Congress of South African Sugar Technologists Association*, **75**, 217–221 (2001).
76. Brown, G. G., Barois, I., and Lavelle, P. Regulation of soil organic matter dynamics and microbial activity in the drilosphere and the role of interactions with other edaphic functional domains. *European Journal of Soil Biology*, **36**(3–4), 177–198 (2000).
77. Kretzschmar, A. Importance of the interaction of soil fauna and microflora for formation of humus and development of organic substance. *Berichte uber Landwirtschaft son der heft*, **206**, 117–126 (1992).
78. Bhatnagar, T. Lombriciens et humification: un aspect nouveau de l'incorporation microbienne d'azote induite par les vers de terre. In: *Biodegradation et Humification*, Pierron and Satreguemines (Eds.). Paris, France, pp. 169–182 (1975).
79. Bhat, J. V. Suitability of experimentation diets for earthworm culture. *Current Science*, **43**, 266–268 (1974).
80. Kale, R. D., Vinayaka, K., Bano, K., and Bagyaraj, D. J. Suitability of neem cake as an additive in earthworm fed and its influence on the establishment of microflora. *Journal of Soil Biology and Ecology*, **6**, 98–103 (1986).
81. Nagar, R., Joshi, N., Dwivedi, S., Khaddar, V. K. A physicochemical and micofloral profile of vermicompost in Tarai region of Himalaya in winter season. *Research on Crops*, **5**, 51–54 (2004).
82. Parthasarathi, K., Ranganathan, L. S., and Zeyer, J. Species specific predation of fungi by *Lampito mauritii* (Kinb) and *Eudrilus eugeniae* (Kinb) reared on press mud. In: Session 3. Earthworms and Vermicomposting, D. J. Bagyaraj (Ed.). *VII National Symposium on Soil Biology and Ecology*, p. 69 (2001).
83. Rajani, B. S., Radhakrishna, D., Ramakrishnaparama, V. R., Kale, R. D., and Balakrishna, A. N. Influence of vermicomposting of urban solid wastes on microbial density and enzyme activities. In: Session 4, Soil Biota as Bioindicators of Soil Health, D. J. Bagyaraj (Ed.). *VII National Symposium on Soil Biology and Ecology*, p. 69 (2001).
84. Raghavendra, Rao B. Assessment of microbial and biochemical quality of urban compost and its impact on soil health. Ph.D. thesis. Bangalore, India, University of Agricultural Sciences (2001).
85. Prakash, M., Jayakumar, M., and Karmegam, N. Physico-chemical characteristics and fungal flora in the casts of the earthworm, Perionyx ceylanensis Mich. reared in Polyalthia longifolia leaf litter. *Journal of Applied Sciences Research*, **4**, 53–57 (2008).
86. Jayakumar, M., Karthikeyan, V., and Karmegam, N. Comparative studies on physico-chemical, microbiological and enzymatic activities of vermicasts of the earthworms, *Eudrilus eugeniae*, *Lampito mauritii* and *Perionyx* ceylanensis cultured in press mud. *International Journal of Applied Agricultural Research*, **4**, 75–85 (2009).
87. Hinton, J. J. and Veiga, M. M. Earthworms as bioindicators of mercury pollution from mining and other industrial activities. *Geochemistry: Exploration, Environment, Analysis*, **2**(3), 269–274 (2002).
88. Mahmoud, H. M. Earthworm (*Lumbricus terrestris*) as indicator of heavy metals in soils. *Online Journal of Veterinary Research*, **11**, 23–37 (2008).
89. Iwai, C. B., Yupin, P., and Noller, B. N. Earthworm: potential bioindicator for monitoring diffuse pollution by agrochemical residues in Thailand. *KKU Research Journal*, **139**, 1081–1088 (2008).
90. Suthar, S. Earthworm communities as bioindicator of arable land management practices: a case study in semiarid region of India. *Ecological Indicators*, **9**(3), 588–594 (2009).
91. Tondoh, J. E., Monin, L. M., Tiho, S., and Csuzdi, C. Can earthworms be used as bio-indicators of land-use perturbations in semi-deciduous forest? *Biology and Fertility of Soils*, **43**(5), 585–592 (2007).
92. Venkateswara, Rao J., Kavitha, P., and Padmanabha, Rao A. Comparative toxicity of tetra ethyl lead and lead oxide to earthworms, *Eisenia fetida* (Savigny). *Environmental Research*, **92**(3), 271–276 (2003).

93. Suthar, S., Singh, S., and Dhawan, S. Earthworms as bioindicator of metals (Zn, Fe, Mn, Cu, Pb, and Cd) in soils: is metal bioaccumulation affected by their ecological category?. *Ecological Engineering*, **32**(2), 99–107 (2008).
94. Sizmur, T. and Hodson, M. E. Do earthworms impact metal mobility and availability in soil?—a review. *Environmental Pollution*, **157**(7), 1981–1989 (2009).
95. Suthar, S. Growth and fecundity of earthworms: *Perionyx excavatus* and Perionyx sansibaricus in cattle waste solids. *The Environmentalist*, **29**(1), 78–84 (2009).
96. Karmegam, N. and Daniel, T. Investigating efficiency of *Lampito mauritii* (Kinberg) and Perionyx ceylanensis Michaelsen for vermicomposting of different types of organic substrates. *The Environmentalist*, **29**(3), 287–300 (2009).
97. Karmegam, N. and Daniel, T. Growth, reproductive biology and life cycle of the vermicomposting earthworm, Perionyx ceylanensis Mich. (Oligochaeta: Megascolecidae). *Bioresource Technology*, **100**(20), 4790–4796 (2009).

8 Urban Green Waste Vermicompost

Swati Pattnaik and M. Vikram Reddy

CONTENTS

8.1 INTRODUCTION

The urban green waste generally comprises of garden or park waste such as grass or flower cuttings and hedge trimmings, domestic and commercial food waste, and vegetable market waste, the latter is generated in large quantities and accumulated in unhygienic way adjacent to vegetable markets emanating unbearable malodor due to lack of proper scientific disposal management particularly in developing countries like India. The vegetable market waste (MW) is the leftover and discarded rotten vegetables, fruits, and flowers in the market. This urban waste can be converted to a potential plant nutrient enriched resource—compost and vermicompost (VC) that can be utilized for sustainable land restoration practices [1]. Vermicomposting is a mesophilic process and is the process of ingestion, digestion, and absorption of organic waste carried out by earthworms followed by excretion of castings through the worm's metabolic system, during which their biological activities enhance the levels of plant nutrients of organic waste [2]. Compost and VC are the end products of aerobic composting process, the later with using earthworms. The VC possessed higher and more soluble level of major nutrients—nitrogen, phosphorus, potassium, calcium, and magnesium [3-5]—compared to the substrate or underlying soil, and normal compost. During the process, the nutrients locked up in the organic waste are changed to simple and more readily available and absorbable forms such as nitrate or ammonium nitrogen, exchangeable phosphorus and soluble potassium, calcium, magnesium in worm's gut [6, 7]. The VC is often considered a supplement to fertilizers and it releases the major and minor nutrients slowly with significant reduction in C/N ratio, synchronizing with the requirement of plants [8].

The vegetable MW as well as floral (*Peltophorum pterocarpum*) waste (FW) were collected and composted using three different earthworm species—*Eisenia fetida*, *Eudrilus eugeniae*, and *Perionyx excavatus* during the present study. These worms have been considered as key agents for organic waste management through the process of vermicomposting [9-13]. The main aim of the present investigation was to know the extent to which vermicomposting and the normal composting of urban green waste may be combined in order to maximize the potentials of both the processes. Earlier, Graziano and Casalicchio [14] have proposed a combination of aerobic composting and vermicomposting to enhance the value of the final products. Frederickson and Knight [15] have showed that vermiculture and anaerobic systems can be combined to enhance organic matter stabilization. The benefits of a combined system to process urban green waste could include effective sanitization and pathogen control due to an initial brief period of thermophilic composting, enhanced rates of stabilization, plus the production of earthworms, and VC [16]. Stabilization of green waste such as yard waste and vegetable waste through the process of composting and vermicomposting has been carried out earlier [16-18]. The present investigation attempted mainly to evaluate the nutrient status of different VC produced by the three earthworm species and that of compost of urban MW and FW in relation to the respective initial substrates, and also to obtain empirical information on the growth and productivity of the three species of earthworms cultured in the two substrates.

8.2 MATERIALS AND METHODS

8.2.1 Methods of Waste Collection

The MW and FW samples each weighing about 125 kg were collected separately in random manner. The MW, both fresh and decomposed, was collected from the main vegetable market of Puducherry which comprised of different leftover putrefied vegetables such as cabbage, tomato, potato, onion, carrot, turnip, brinjal, and leafy vegetables; the FW was obtained from the *P. pterocarpum* (Family-Fabaceae and Subfamily Caesalpinioideae), a widely appreciated shade tree and a reclamation plant with dense spreading crown, and planted along the roadsides in the Pondicherry University campus. These wastes were characterized by segregating and discarding the non-biodegradable fraction, and the biodegradable component was used for the experiment. Five samples of each waste were taken for experimentation and analyses.

8.2.2 Sample Processing—Pre-composting

The collected MW and FW were air dried separately spreading over a polythene sheet for 48 hr. The air-dried samples were pre-composted for 3 weeks before putting into vermicomposting and composting process. Pre-composting is the pre processed and pretreated practice of raw waste. The waste materials, in the pre-composting process were decomposed aerobically by the active role of bacteria due to which temperature raised up to 60°C. As such a high temperature was lethal for earthworm survival; the thermal stabilization was done prior to introduction of earthworms into the substrate. When the temperature of the pre-composted substrate diminished to 25°C, adult earthworms with well-defined clitella belonging to the three species namely, *E. eugeniae*, *E. fetida*, and *P. excavatus* were introduced on the pre-composted material filled in each set of earthen pots (The earthworms were collected from a local vermiculture unit at Lake Estate of Auorbindo Ashram, Puducherry, India).

8.2.3 Experimental Design

In each pot 5 kg of the substrate mixed with cow dung in ratio were taken for vermicomposting and composting. A total of four sets of earthen pots each set comprising six replicates was taken for each waste, of which three sets were used for vermicomposting with each set using one species of earthworm and the fourth set was used for normal composting that is, without using any earthworm. Three species of earthworms, each of fifty adult individuals, were introduced on the top of the pre-composted substrate in each of the three sets of pots keeping aside the fourth set for composting without earthworms. All the pots were covered on the top by jute cloth cover and wire mesh to prevent and protect the earthworms from the predators—centipedes, moles, and shrews. Small holes were drilled at the bottom of each pot, which was filled with small stones up to a height of 5 cm for air circulation and good drainage. The processes of vermicomposting and composting were carried out for a period of 60 days. The temperature and moisture content were maintained by sprinkling adequate quantity of water at frequent intervals. The harvesting of VC and compost, total earthworm biomass, individual body weight, total numbers of juveniles, adults, and cocoons were carried out, and the mortality rates of the three earthworm species were calculated after 60 days, at the end of the experiment.

8.2.4 Physico-chemical Analyses

The homogenized subsamples of each substrate material and their respective compost and VC samples (on the basis of 100 g dry weight) were collected undestructively at 0 (i.e., substrate), 15, 30, 45, and 60 days from each replicate pot and compound samples were made which were processed for analyses of OC and major nutrients—total nitrogen (N), available phosphorus (P), exchangeable potassium (K), calcium (Ca), and magnesium (Mg). The temperature (°C), moisture (%), pH, and electrical conductivity (EC) were recorded for the substrate and during the vermicomposting and composting processes. Temperature was noted daily using a thermometer, and moisture content was measured gravimetrically. The pH and EC of samples were recorded by a digital pH meter and conductivity meter, respectively. The OC of the samples was measured by Walkey-Black method [19]; the N was estimated by the Kjeldahl method [20], and the P and K contents of the samples were analyzed by calorimetric method [21] and flame photometric method [22], respectively. The Ca and Mg contents of the samples were also analyzed using atomic absorption spectrophotometer (GBC make) [20]. The ratio was calculated from the measured values of C and N.

8.2.5 Statistical Analysis

Two-way analysis of variance (ANOVA) was computed using statistical package for the social sciences (SPSS) (version No. 10) to test the level of significance of difference between the VC produced by the three species earthworms and compost samples with respect to nutrient parameters.

8.3 DISCUSSION AND RESULTS

8.3.1 Growth and Productivity of Earthworms

The growth parameters of three earthworm species cultured in MW and FW showed that the length increased by 23.3% in *E. eugeniae*, 43.7% in *E. fetida*, and 85.0% in *P. excavatus grown* in MW, while it increased by 29.3% in *E. eugeniae*, 53.7% in *E. fetida*, and 122.5% in *P. excavatus* grown in FW, whereas the net individual weight gained by each of the three species was 200.0, 261.2, and 525.8% in MW and 265.7, 432.8, and 906.4% in FW respectively, at the end of the experiment (Table 1). The net individual weight gain and total biomass gain were higher in *P. excavatus* than that of *E. fetida*, and *E. eugeniae*. The total biomass gain was found 1456.6 and 2171.4% by E. eugeniae, 2095.2 and 3465.1% by *E. fetida*, and 4303.9 and 7272.3% by *P. excavatus* in MW and FW respectively, at the end of the vermicomposting process. Cocoon production rate was higher in *P. excavatus* than that of *E. eugeniae* and *E. fetida*. The number of worms produced per cocoon was 28.9 and 71.0% higher in *E. fetida* than that of *E. eugeniae* and P. excavatus, respectively, while the number of cocoons collected at the end of the experiment was more in *P. excavatus* by 245.6% than that of *E. eugeniae* and 286.3% than that of *E. fetida* in MW; and by 186.8% and 194.6% than that of *E. eugeniae* and *E. fetida* in FW, respectively. The number of juveniles collected was 83.3% higher in *P. excavatus* than that of *E. eugeniae* and 50.5% than that of *E. fetida* in MW, whereas the increase was 74.2% in *E. eugeniae* and 30.6% in *E. fetida*. Adult earthworm number was higher in *P. excavatus* than that of *E eugeniae*, and *E. fetida* by 35.8 and 15.8% in MW, and 16.8 and 9.4% in FW, respectively. The production of cocoons, juveniles, and adults of all the three species was higher in FW than that of MW, which indicated the former waste material as a better substrate for the earthworms. The mortality rate of the *P. excavatus* was 900% higher than that of *E eugeniae* and 400% higher than that of E. *fetida* grown in MW, while it was higher by 1100 and 650% than *E eugeniae* and *E. fetida* grown in FW, respectively.

The mean individual length and live weight, mean growth rate of an individual (mg/day), individual and total biomass gain, reproduction rate (cocoon worm^{-1} day^{-1}), fecundity rate (worm cocoon $^{-1}$ day^{-1}), total cocoon, juveniles and adult numbers, and

TABLE 1 Growth parameters of three earthworm species during the process of vermicomposting of MW and FW.

Earthworm growth parameters	***E. eugeniae***		***E. fetida***		P. excavates	
	MW	**FW**	**MW**	**FW**	**MW**	**FW**
Av. Individual length:						
Initial (cm)	15.0 ± 0.02	15.0 ± 0.02	8 ± 0.01	8 ± 0.01	4 ± 0.0 2	4 ± 0.02
Final (cm)	18.5 ± 0.04	19.4 ± 0.05	11.5 ± 0.05	12.3 ± 0.04	7.4 ± 0.06	8.9 ± 0.07
Av. Individual weight						
Initial (gm)	3.5 ± 0.02	3.5 ± 0.02	0.67 ± 0.01	0.67 ± 0.01	0.31 ± 0.0 2	0.31 ± 0.02
Final (gm)	10.5 ± 0.05	12.8 ± 0.1	2.42 ± 0.01	3.57 ± 0.02	1.94 ± 0.04	3.12 ± 0.03
Av. Total biomass						
Initial (gm)	175 ± 0.06	175 ± 0.06	33.5 ± 0.05	33.5 ± 0.05	15.5 ± 0.05	15.5 ± 0.05
Final (gm)	2724 ± 0.2	3975 ± 0.4	735.4 ± 0.06	1194.3 ± 0.0	682.6 ± 0.0 4	1142.7 ± 0.0
Av. Cocoon production rate	0.51 ± 0.006	0.51 ± 0.006	0.5 ± 0.003	0.5 ± 0.003	2.7 ± 0.00 1	2.7 ± 0.001
Av. Worm number per cocoon	2.7 ± 0.09	2.7 ± 0.09	3.8 ± 0.01	3.8 ± 0.01	1.1 ± 0.03	1.1 ± 0.03

TABLE 1 *(Continued)*

Earthworm growth parameters	***E. eugeniae***		***E. fetida***		P. excavates	
	MW	**FW**	**MW**	**FW**	**MW**	**FW**
Av. Cocoon number at the end	57 ± 0.3	76 ± 0.07	51 ± 0.05	74 ± 0.06	197 ± 0.05	218 ± 0.03
Av. Juvenile number at the end	78 ± 0.08	93 ± 0.06	95 ± 0.07	124 ± 0.05	143 ± 0.06	162 ± 0.06
Av. Adult number at the end	254 ± 0.04	310 ± 0.04	298 ± 0.08	331 ± 0.06	345 ± 0.04	362 ± 0.09
Av. Mortality rate	0.03 ± 0.002	0.05 ± 0.003	0.06 ± 0.004	0.08 ± 0.005	0.3 ± 0.02	0.6 ± 0.03

mortality rate in the present study varied across different treatments. The worms when introduced into wastes showed an increased growth rate and reproduction activities [1]. The increase in body weight of all three earthworm species was noted in both the substrates during vermicomposting process, which could be due to the substrate quality or could be related to fluctuating environmental conditions [23-25]. The readily available nutrients in MW and FW enhanced the feeding activity of the worms, showing their increase in biomass [1]. Interestingly, cocoon production rate was higher in *P. excavatus*, whereas the number of worms per cocoon was higher in *E. fetida* compared to other species. The indigenous species, *P. excavates*, exhibited better growth and reproduction performance compared to the other two exotic species [26]. The higher numbers of cocoons, juveniles, and adults collected from the VC processed by *P. excavatus*, were probably because its indigenous nature being acclimatized to the abiotic environmental conditions extremely well compared to other species. The difference in worm mortality among the three species could be related to the species-specific composting behavior or to specific tolerance nature of earthworm according to the changing microenvironmental conditions in composting subsystem [1]. Moreover, the growth rate difference between the three species was probably due to the species-specific growth patterns or could be related to the feed quality and preferences by individual species of earthworm [1].

8.3.2 Waste Stabilization

The reduction in bulk dry mass of both the substrates—MW and FW, the range of temperature, moisture content, pH, EC of the substrate, compost and VC presented in Table 2 depicted that higher mass reduction of MW was recorded in the VC processed by *E. eugeniae* (75%), followed by that of *E. fetida* (63%), and *P. excavatus* (50%) compared to that of compost (26%), whereas the mass reduction was higher 83% in VC produced by *E. eugeniae*, 67% by that of E. fetida, 56% in that of *P. excavates*, and 30% in sole compost than that of FW. The marked stabilization of both the substrates due to vermicomposting process was higher in the VC processed by *E. eugeniae* compared to that of other two and the compost. The FW and its VC and composts were found to be more stabilized than that of MW.

The pre-composting because of its thermophilic nature prior to vermicomposting helped in mass reduction and pathogen reduction [27]. It was found that the bulk (dry) mass reduction and stabilization of both the wastes during present study through vermicomposting process were significant [2, 27]; the vermicomposting may also be known as vermistabilization [28]. The cow dung used as the inoculant in the vermicomposting process enhanced the quality of feeding resource attracting the earthworms and accelerated the breakdown of wastes resulting in the reduction of ratio by increasing certain nutrients [1, 29-31].

8.3.3 Physical State of MW and FW during Vermicomposting and Composting Processes

The physical characteristics recorded during the period of this study presented in Table 2 were conducive for vermicomposting process [6, 32]. The temperature ranged from 22.3 to 29.8°C and was lower by 19.1 and 15.8% in the VC processed by *E. eugeniae*,

TABLE 2 Weight and other different physical parameters of the substrates—MW and FW—and their respective compost and VC of three earthworm species (Mean ± sd).

Parameters	0 days		60 days							
			Vermicompost						Compost	
			E. eugeniae		E. fetida		P. excavatus			
	MW	FW	MW	FW	MW	FW	MW	FW	MW	FW
Weight (kg)	5.00 ± 0.005	5.00 ± 0.01	1.25 ± 0.03	0.85 ± 0.05	1.85 ± 0.04	1.65 ± 0.04	2.5 ± 0.03	2.2 ± 0.01	3.7 ± 0.009	3.5 ± 0.008
Temperature (°c)	29.8 ± 0.06	26.5 ± 0.05	24.1 ± 0.04	22.3 ± 0.03	24.2 ± 0.05	23.4 ± 0.03	24.4 ± 0.04	23.5 ± 0.05	2 4.7 ± 0.05	23.9 ± 0.02
Moisture content (%)	55.73 ± 0.08	34.62 ± 0.03	65.2 ± 0.03	60.8 ± 0.08	64.72 ± 0.03	59.67 ± 0.02	64.04 ± 0.01	58.68 ± 0.05	63.11 ± 0.04	57.49 ± 0.05
pH	6.31 ± 0.07	6.84 ± 0.04	7.12 ± 0.02	7.37 ± 0.02	7.08 ± 0.01	7.28 ± 0.03	6.95 ± 0.02	6.89 ± 0.04	6.87 ± 0.03	6.79 ± 0.04
Electric conductivity (mhos/cm)	495.5 ± 0.04	152.2 ± 0.02	3354.4 ± 0.02	532.5 ± 0.03	2716.7 ± 0.07	466.3 ± 0.03	1983.2 ± 0.06	415.7 ± 0.02	1789.3 ± 0.07	363.5 ± 0.01

by 18.8 and 11.7% in that of *E. fetida*, by 18.1 and 11.3% in that of *P. excavates*, and by 17.1 and 9.8% in compost than that of initial substrate of MW and FW, respectively. The moisture content of VC of *E. eugeniae* varied by 17.0 and 75.6%, by 16.1 and 72.4% in that of *E. fetida*, by 14.9 and 69.5% in that of *P. excavates*, and by 13.2 and 66.1% in the compost than that of initial MW and FW, respectively. The pH ranged from 6.31 to 7.37 and increased by 12.8, 12.2, 10.1, and 8.9% than that of MW; and 7.7, 6.4, 2.1 and 0.7% than that of FW, in VC of *E. eugeniae, E. fetida, P. excavatus*, and compost, respectively. The EC of VC ranged from 152.2 to 3354.4 mhos/cm and increased EC noted in VC processed by *E. eugeniae, E. fetida, P. excavatus* and in compost was 577.0, 448.3, 300.2, and 261.1% more than that of MW, and was 249.9, 206.4, 173.1, and 138.8% more than that of FW, respectively, at the end of composting process. Temperature, moisture content, and EC were more and pH was less in MW compared to that of FW.

Temperature

At the start of the experiment, the temperature of the substrate was high and then decreased gradually as the composting process progressed. The heat released by the oxidative action of intensive microbial activity on the organic matter resulted in the rise in temperature during the first mesophilic phase of composting process [33]. The temperature of the following thermophilic phase rose up above 40°C reaching about 60°C when most of the organic matter was degraded with the help of thermophilic bacteria and fungi, consequently depleting most of the oxygen. The thermophilic phase was followed by cooling phase, when compost maturation stage occurred and compost temperature dropped to that of the ambient [34]. Then, the decreasing trend of temperature with the progress of composting process occurred which was probably due to the decreased bacterial activity. It may also be attributable to regular sprinkling of water.

Moisture Content

Moisture content ranged from 50 to 70% [35]. Edwards and Bater [36] reported that optimum moisture content for growth of earthworms—*E. fetida, E. eugeniae,* and *P. excavatus*—was 85% in organic waste management. The rate of mineralization and decomposition becomes faster with the optimum moisture content [37]. According to Liang et al. [38], the moisture content of 60–70% was proved having maximal microbial activity, while 50% moisture content was the minimal requirement for rapid rise in microbial activity. The VC samples during the present study showed higher moisture content than the compost and substrate which may be due to their high absorption capacity, and may also be because of assimilation rate by microbial population indicating the higher rate of degradation of waste by earthworms. Relatively highest moisture content of VC produced by *E. eugeniae* followed by that of *E. fetida* and *P. excavatus* implied greater palatability of the substrate by the species.

pH

It was neutral being around 7 and increased gradually from substrate to compost to VC [35, 39]. The near-neutral pH of VC may be attributed by the secretion of NH_4^+ ions that reduce the pool of H^+ ions [40] and the activity of calciferous glands in earthworms containing carbonic anhydrase that catalyzes the fixation of CO_2 as $CaCO_3$, thereby preventing the fall in pH [9]. The increased trend of pH in the VC and compost samples is in consistence with the findings of Tripathi and Bhardwaj [41] and Loh et al. [42] which was due to higher mineralization, whereas the present findings are in contradiction to the findings of Suthar and Singh [1], Haimi and Huhta [40], and Ndegwa et al. [43] who reported lower pH. The increased pH during the process was probably due to the degradation of short-chained fatty acids and ammonification of organic N [44-46]. Fares et al. [47] found the increased pH at the end of the composting process, which was attributed to progressive utilization of organic acids and increase in mineral constituents of waste.

EC

The increased EC during the period of the composting and vermicomposting processes is in consistence with that of earlier workers [48, 49] which was probably due to the degradation of organic matter releasing minerals such as exchangeable Ca, Mg, K, and P in the available forms, that is in the form of cation in the VC and compost [44, 46].

8.3.4 Nutrients in MW and FW and their VC and Compost

It was found that the N was 0.45% in MW and 0.17% in FW, P was 0.25% in MW and 0.11% in FW, K was 0.18% in MW and 0.02% in FW, Ca was 0.62% in MW and 0.07% in FW, and Mg was 0.17% in MW and 0.04% in FW, while the content of OC was 79.6% in MW and 42.9% in FW (Figures 1 and 2).

The present study revealed that all VC prepared from their respective organic wastes possessed considerably higher levels of major nutrients—N, P, K, Ca, and Mg compared to that of the substrates [31, 50]. The increase in the nutrients and decrease in OC, C/N ratio, and C/P ratios in the VC, are in consistence with the findings of earlier investigators [25, 26]. Moreover, comparing the nutrient contents of VC with that of compost, VC possessed significantly higher concentrations of nutrients than that of compost ($P < 0.05$), which was probably due to the coupled effect of earthworm activity as well as a shorter thermophilic phase [51, 52], making the plant availability of most the nutrients higher in vermicomposting than that of composting process [3, 53, 54].

8.3.5 Temporal Variation in Nutrients

In the present study the percentage of OC decreased (Figure 1(a)) and that of N increased (Figure 1(b)), while the percentage of P (Figure 1(c)) and K (Figure 1(d)), and that of Ca (Figure 2(a)) and Mg (Figure 2(b)) also increased gradually in all the three VC and in the sole compost as the composting process progressed from 15 days to 60 days. Interestingly, the C/N ratio (Figure 2(c)) and C/P ratio (Figure 2(d)) in all the samples of VC and compost declined at the end of the experiment (i.e., after 60 days of processing). The nutrient contents showed significant temporal variation in VC and compost of both the substrates, that is MW (Table 3) and FW (Table 4).

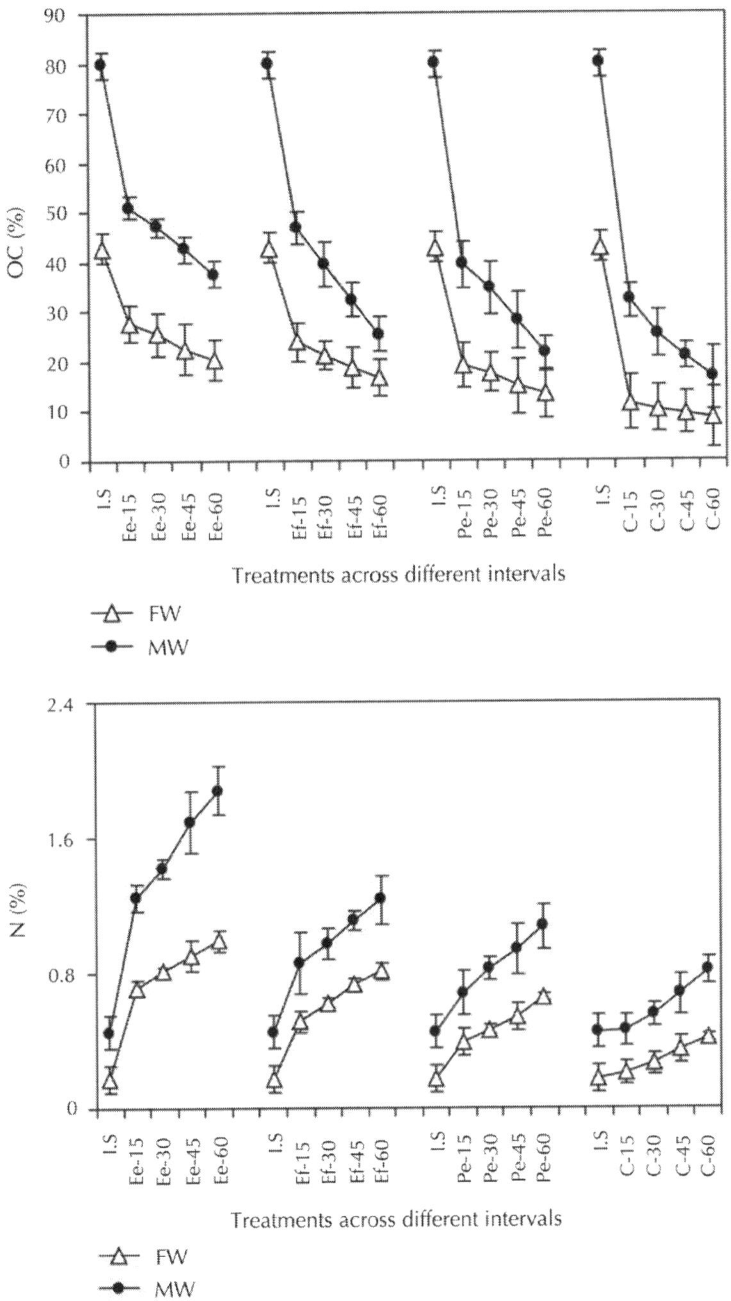

FIGURE 1 *(Continued)*

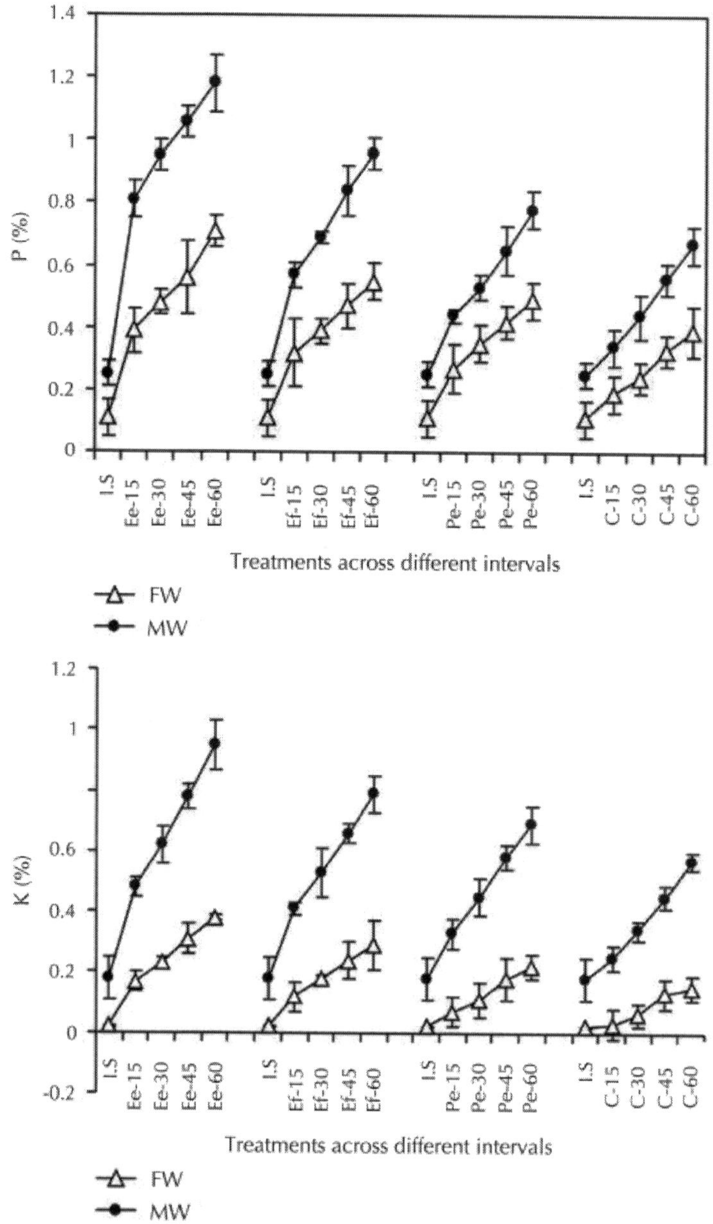

FIGURE 1 Major nutrients—OC, N, P, and K (%) of VC of three different species of earthworms—*Eudrilus eugeniae* (*Ee*) at 15 days (Ee-15), 30 days (Ee-30), 45 days (Ee-45), and 60 days (Ee-60); *Eisenia fetida* (*Ef*) at 15 days (Ef-15), 30 days (Ef-30), 45 days (Ef-45), and 60 days (Ef-60); *Perionyx excavatus* (*Pe*) at 15 days (Pe-15), 30 days (Pe-30), 45 days (Pe-45), and 60 days (Pe-60); and Compost (C) at 15 days (C-15), 30 days (C-30), 45 days (C-45), and 60 days (C-60) produced from FW and MW. (a) OC, (b) N, (c) P, and (d) K.

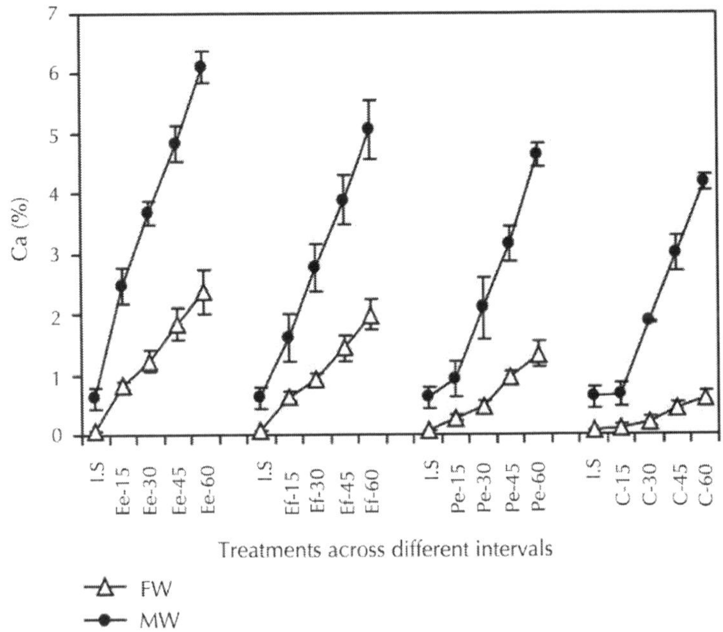

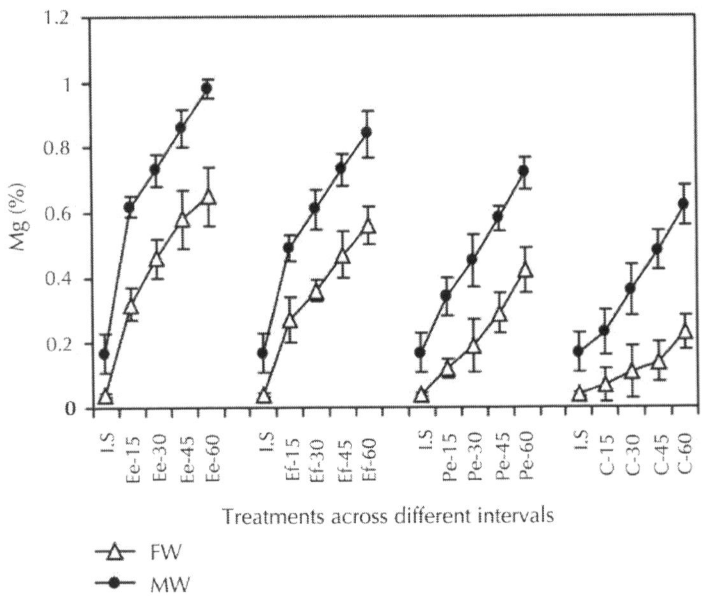

FIGURE 2 *(Continued)*

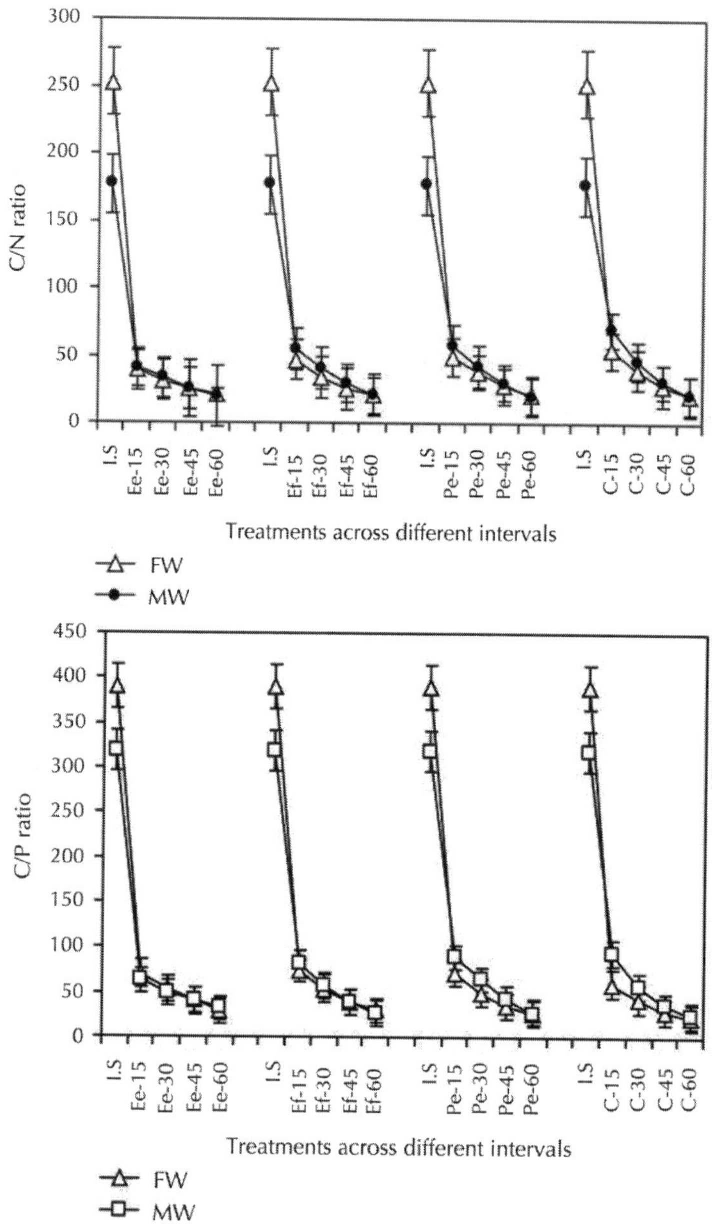

FIGURE 2 Major nutrients—Ca and Mg (%), C/N ratio and C/P ratio of VC of three different species of earthworms—*Eudrilus eugeniae* (*Ee*) at 15 days (Ee-15), 30 days (Ee-30), 45 days (Ee-45), and 60 days (Ee-60); *Eisenia fetida* (*Ef*) at 15 days (Ef-15), 30 days (Ef-30), 45 days (Ef-45), and 60 days (Ef-60); *Perionyx excavatus* (Pe) at 15 days (Pe-15), 30 days (Pe-30), 45 days (Pe-45), and 60 days (Pe-60); and Compost (C) at 15 days (C-15), 30 days (C-30), 45 days (C-45), and 60 days (C-60) produced from FW and MW. (a) Ca, (b) Mg, (c) C/N ratio, (d) C/P ratio.

TABLE 3 The ANOVA of different nutrients of VC produced by three species of earthworms and compost (Treatments) of MW across different time intervals.

Source of Variation	SS	df	MS	F
OC				
Time Intervals	648.6706	3	216.2235	83.74185**
Treatments	923.0771	3	307.6924	119.1671**
Error	23.23823	9	2.582025	
N				
Time Intervals	0.431569	3	0.143856	38.0167**
Treatments	1.881169	3	0.627056	165.7113**
Error	0.034056	9	0.003784	
C/N Ratio				
Time Intervals	2834.197	3	944.7322	36.40393**
Treatments	301.5306	3	100.5102	3.87302**
Error	233.5624	9	25.95138	
P				
Time Intervals	0.286919	3	0.09564	418.6049**
Treatments	0.568369	3	0.189456	829.231**
Error	0.002056	9	0.000228	
C/P Ratio				
Time Intervals	6752.972	3	2250.991	39.35673**
Treatments	225.6022	3	75.20074	1.314823**
Error	514.751	9	57.19456	
K				
Time Intervals	0.32795	3	0.109317	141.5612**

TABLE 3 *(Continued)*

Source of Variation	SS	df	MS	F
Treatments	0.2005	3	0.066833	86.54676**
Error	0.00695	9	0.000772	
Ca				
Time Intervals	28.18897	3	9.396323	2027.679**
Treatments	8.064019	3	2.688006	580.0583**
Error	0.041706	9	0.004634	
Mg				
Time Intervals	0.30515	3	0.101717	1220.6**
Treatments	0.3242	3	0.108067	1296.8**
Error	0.00075	9	8.33E-05	

Level of significance: $^{**}P < 0.001$

TABLE 4 The ANOVA of different nutrients of VC produced by three species of earthworms and compost (Treatments) of FW across different time intervals.

Source of Variation	SS	df	MS	F
OC				
Time Intervals	76.11592	3	25.37197	28.39579**
Treatments	426.3413	3	142.1138	159.0508**
Error	8.041606	9	0.893512	
N				
Time Intervals	0.151425	3	0.050475	118.7647**
Treatments	0.649525	3	0.216508	509.4314**
Error	0.003825	9	0.000425	
C/N Ratio				
Time Intervals	1629.242	3	543.0806	67.49946**

TABLE 4 *(Continued)*

Source of Variation	SS	df	MS	F
Treatments	103.9289	3	34.64297	4.305774**
Error	72.41133	9	8.045703	
P				
Time Intervals	0.130719	3	0.043573	78.33333**
Treatments	0.127569	3	0.042523	76.44569**
Error	0.005006	9	0.000556	
C/P Ratio				
Time Intervals	4035.872	3	1345.291	636.1514**
Treatments	295.7819	3	98.59395	46.6224**
Error	19.0326	9	2.114733	
K				
Time Intervals	0.062619	3	0.020873	81.45528**
Treatments	0.072769	3	0.024256	94.65854**
Error	0.002306	9	0.000256	
Ca				
Time Intervals	2.90885	3	0.969617	23.82676**
Treatments	3.5505	3	1.1835	29.08259**
Error	0.36625	9	0.040694	
Mg				
Time Intervals	0.1621	3	0.054033	35.62637**
Treatments	0.31855	3	0.106183	70.01099**
Error	0.01365	9	0.001517	

Level of significance: **P< 0.001

The VC of MW produced by *E. eugeniae* showed 177.8, 224.0, 166.7, 296.7, and 264.7% increase after 15 days of processing and 317.8, 372.0, 427.8, 887.1, and 476.5% increase after 60 days of processing in N, P, K, Ca, and Mg compared to that of the substrate, respectively, whereas it decreased by 35.9 and 52.8% after 15 and 60 days, respectively in OC; whereas that of *E. fetida* increased by 91.1, 128.0, 127.8, 161.3, and 188.2%; and 173.3, 284.0, 338.9, 716.1, and 394.1% while decreased by 41.2% and 68.1% after 15 and 60 days of processing, respectively. The N, P, K, Ca, and Mg contents in VC produced by *P. excavatus* increased by 51.1, 76.0, 83.3, 50.0, and 100.0%, respectively and the OC decreased by 50.5%, at 15 days of processing; whereas the increase was 137.8, 212.0, 283.3, 648.4, and 323.5% and the decrease was 73.1% at 60 days of processing, respectively. In compost, the increase was relatively less and was 2.2, 36.0, 38.9, 4.8, and 35.3% and 80.0, 168.0, 216.7, 572.6, and 264.7% in N, P, K, Ca, and Mg, respectively and its decrease in OC was 59.7, and 79.1% compared to that of substrate after 15 and 60 days of composting process, respectively. The C/N ratio reduction was 76.9, 69.2, 67.3, and 60.6% after 15 days of processing and 88.7, 88.4, 88.7, and 88.4% after 60 days while the C/P ratio reduction was respectively 80.2, 74.2, 71.9, and 70.4% at 15 days of processing and 90.0, 91.7, 91.2, and 92.2% at 60 days of processing in the VC produced by *E. eugeniae, E. fetida, P. excavatus*, and in sole compost compared to that of the substrate.

The VC of FW produced by *E. eugeniae* increased by 317.6, 254.5, 750.0, 1057.1, and 700.0% after 15 days of processing and 482.3, 545.4, 1800.0, 3285.7, and 1525.0% after 60 days of processing in N, P, K, Ca, and Mg, respectively compared to the substrate, whereas it decreased by 35.7 and 52.6% after 15 and 60 days, respectively in OC, while that of *E. fetida* increased by 200.0, 190.9, 500.0, 814.3 and 575.0% and 376.5, 400.0, 1350.0, 2728.6, and 1300%; while decreased by 44.3, and 60.9% after 15 and 60 days of processing, respectively. The N, P, K, Ca, and Mg contents in VC produced by *P. excavatus* increased by 129.4, 145.4, 250.0, 285.7, and 200.0%, respectively, and the OC decreased by 55.2% at 15 days of processing, whereas the increase was 282.3, 345.4, 1000.0, 1785.7, and 950.0% and the decrease was 68.5% at 60 days of processing, respectively. In compost, there was less increase and was 23.5, 72.7, 50.0, 28.6, and 75.0% and 141.2, 254.5, 650.0, 742.8, and 475.0% in N, P, K, Ca, and Mg, respectively, and its decrease in OC was 73.4, and 79.9% after 15 and 60 days, respectively of composting process compared to that of substrate. The C/N ratio reduction was 84.6, 81.4, 80.5 and 78.5% after 15 days of processing and 91.9, 91.8, 91.7, and 91.7% after 60 days while the C/P ratio reduction was respectively 81.9, 80.8, 81.7, and 84.6% at 15 days of processing and 92.6, 92.2, 92.9, and 94.3% at 60 days of processing in the VC produced by *E. eugeniae, E. fetida, P. excavatus* ,and in sole compost more than that of the substrate.

The considerable enrichment of nutrients of the vermicompost of the three species of earthworms—*E. eugeniae, E. fetida,* and *P. excavatus*—compared to that of composts of substrates, that is MW and FW, were consistent with the findings of earlier reports [2, 25, 26, 30]. At the end of the experiment, the increase in OC, N, P, K, Ca, and Mg was 55.8, 56.9, 43.2, 40.0, 31.8, and 36.7% in the VC of MW and 57.7, 58.6, 45.1, 60.5, 75.1, and 64.6% in that of FW produced by *E. eugeniae*; 34.5, 34.1, 30.2, 27.8, 17.6, and 26.2% in that of MW and 48.6, 49.4, 29.1, 48.3, 70.2, and 58.9% in that of FW produced by *E. fetida*; and 22.5, 24.3, 14.1, 17.4, 10.1, and 13.9% in that of MW and 36.3, 36.9, 20.4, 31.8, 55.3, and 45.2% in that of FW produced by *P. excavatus*, compared to that of sole compost, respectively.

The nutrients and OC were found higher in MW compared to that of FW, which was most probably because of mosaic nature of the MW. In all the VC and compost of the present study the nutrients increased and OC, C/N ratios, and C/P ratio decreased significantly with the passage of time (from 0 to 15, 30, 45, and 60 days), from the substrate (organic waste) to compost and VC, respectively [2]. The present findings are in agreement with the findings of earlier workers: Nagavallemma et al. [35], Uthaiah [55], Muthukumarasamy et al. [56], Parthasarathi and Ranganathan [57], and Khwairakpam and Bhargava [58]. The waste materials ingested by the earthworms undergo physical decomposition and biochemical changes contributed by the enzymatic and enteric microbial activities while passing through the earthworm gut due to the grinding action of the muscular gizzard releasing the nutrients in the form of microbial metabolites enriching the feed residue with plant nutrients and growth promoting substances in an assimilated form, which is excreted in the form of vermicast [31, 59].

Comparing the nutrients of VC produced by the three earthworm species (*E. eugeniae, E. fetida*, and *P. excavatus*), it was found that the VC of *E. eugeniae* possessed significantly higher concentrations of the nutrients followed by E. fetida and P. excavates, and the sole compost, in the order of E. *Eugeniae* > *E. fetida* > *P. excavatus* > compost, which may indicate that the earthworm is more efficient in recovering nutrients from the waste through vermicomposting process [2, 60]. However, the findings of Sangwan et al. [61], in contrast to the present findings, reported decrease in potassium content in the VC produced by *E. fetida* compared to that of the substrate. Khwairakpam and Bhargava [58] compared the VC of sewage sludge processed by these three earthworm species in order to report the suitability of worm species for composting. Reddy and Okhura [5] have assessed the VC produced by different earthworm species—*Perionyx excavatus, Octohaetona phillotti*, and *Octonachaeta rosea* using the rice straw as substrate and found that VC produced *P. excavatus* possessed higher concentration of nutrients than that of *O. rosea* and *O. phillotti*.

Further, it was found that the OC, N, P, K, Ca, Mg was 85.4, 164.7, 127.3, 800.0, 785.7, and 325.0%, respectively increased in MW than that of FW, and the nutrients were also significantly higher in the VC and compost of MW than that of FW ($P< 0,05$). The VC of MW produced by *E. eugeniae* showed 84.9, 76.0, 107.7, 182.3, 203.7, and 93.7% increase at 15 days and 84.8, 89.9, 66.2, 150.0, 158.2, and 50.8% increase at 60 days of processing in OC, N, P, K, Ca, Mg than that of FW; whereas the increase was 95.8, 68.6, 78.1, 241.7, 153.1, and 81.5% at 15 days of composting and 51.4, 51.8, 74.5, 172.4, 155.6 and 50.0% at 60 days in the VC produced by E. fetida, and the increase of OC, N, P, K, Ca, and Mg (Figures 1 and 2) in VC of MW produced by *P. excavatus* that of FW was 104.7, 74.4, 62.9, 371.4, 244.4, and 183.3% and 58.5, 64.6, 59.2, 213.6, 251.5, and 71.4% after 15 and 60 days of processing, respectively. The compost of MW was higher by 181.1 and 92.9% in OC, 119.0 and 97.6% in N, 78.9 and 71.8% in P, 733.3 and 280.0% in K, 622.2 and 606.8% in Ca, and 228.6 and 169.6% in Mg after 15 and 60 days of processing, respectively compared to that of FW.

Total N

The total N content of VC of the tree earthworm species was higher than that of compost and substrate. The increasing trend of N in the VC produced by the earthworm species in the present study corroborated with the findings of earlier reports [62, 63]. The enhancement of N in VC was probably due to mineralization of the organic matter containing proteins [3, 8] and conversion of ammonium nitrogen into nitrate [1, 64]. Earthworms can boost the nitrogen levels of the substrate during digestion in their gut adding their nitrogenous excretory products, mucus, body fluid, enzymes, and even through the decaying dead tissues of worms in vermicomposting subsystem [25]. The VC prepared by all the three earthworm species showed a substantial difference in total N content, which could be attributed directly to the species-specific feeding preference of individual earthworm species and indirectly to mutualistic relationship between ingested microorganisms and intestinal mucus [1].

OC

Total OC decreased with the passage of time during vermicomposting and composting processes in both the substrates. These findings are in consistence with those of earlier authors [12, 46]. The OC is lost as CO_2 through microbial respiration and mineralization of organic matter causing increase in total N [65]. Part of the carbon in the decomposing residues released as CO_2 and a part was assimilated by the microbial biomass [11, 66, 67]; microorganisms used the carbon as a source of energy decomposing the organic matter. The reduction was higher in vermicomposting compared to the ordinary composting process, which may be due to the fact that earthworms have higher assimilating capacity. The difference between the carbon loss of the VC processed by *E. eugeniae, E. fetida*, and *P. excavatus* could be due to the species-specific differences in their mineralization efficiency of OC.

C/N Ratio

The C/N ratios of VC of three earthworm species were around; such ratios make nutrients easily available to the plants. Plants cannot assimilate mineral N unless the C/N ratio is about, and this ratio is also an indicative of acceptable maturity of compost [68]. The C/N ratio of the substrate material reflects the organic waste mineralization and stabilization during the process of composting or vermicomposting. Higher C/N ratio indicates slow degradation of substrate [69], and the lower the C/N ratio, the higher is the efficiency level of mineralization by the species. Lower C/N ratio in VC produced by *E. eugeniae* implied that this species enhanced the organic matter mineralization more efficiently than *E. fetida* and *P. excavatus* [1, 60]. The loss of carbon through microbial respiration and mineralization and simultaneous addition of N by worms in the form of mucus and nitrogenous excretory material lowered the C/N ratio of the substrate [25, 70-72].

P

The total P was higher in the VC harvested at the end of the experiment compared to that of the initial substrate [8, 25, 73]. The enhanced P level in VC suggests P mineralization during the process. The worms during vermicomposting converted the insoluble P into soluble forms with the help of P-solubilizing microorganisms through phosphatases present in the gut, making it more available to plants [1, 60, 74]. This was buttressed by increased trend of EC showing enhancement of exchangeable soluble salts in VC of all the three earthworm species.

K

Vermicomposting proved to be an efficient process for recovering higher K from organic waste [1, 25, 73]. The present findings corroborated to those of Delgado et al. [75], who demonstrated that higher K concentration in the end product prepared from sewage sludge. The increase in K of the VC in relation to that of the simple compost and substrate was probably because of physical decomposition of organic matter of waste due to biological grinding during passage through the gut, coupled with enzymatic activity in worm's gut, which may have caused its increase [76]. The microorganisms present in the worm's gut probably converted insoluble K into the soluble form by producing microbial enzymes [48].

Ca and Mg

The higher Ca content in VC compared to that of compost and substrate is attributable to the catalytic activity of carbonic anhydrase present in calciferous glands of earthworms generating $CaCO_3$ on the fixation of CO_2 [60]. The higher concentration of Mg in VC reported in present study was also in consistence with the findings of earlier workers [60, 77].

8.4 CONCLUSION

It is concluded that among the three species, the indigenous species, *P. excavates*, exhibited better growth and reproduction performance compared to the other two exotic species. *E. eugeniae* was more efficient in bioconversion of urban green waste into nutrient rich VC compared to *E. fetida* and *P. excavatus*; the VC produced by *E. eugeniae* possessed higher nutrients—N, P, K, Ca and Mg—compared to that of *E. fetida* and *P. excavatus*. The VC produced by all the earthworm species showed higher contents of nutrients compared to that of the sole compost as well as substrates—the green waste (vegetable MW and FW). Moreover, the VC and compost of vegetable MW possessed higher nutrient contents probably because it comprised of a mosaic of materials compared to that of floral waste. Thus, vermicomposting was proved to be a better technology than that of sole composting and may be preferred for the management and nutrient recovery from the urban waste such as MW and FW.

KEYWORDS

- **Floral waste**
- **Market waste**
- **Mesophilic process**
- **Urban green waste**
- **Vermicomposting**

ACKNOWLEDGMENT

The University Grants Commission (New Delhi) provided grants in the form of a Major Research Project for this research, which covered a project fellowship to the first author The earthworm species were procured from the Vermiculture unit of Lake Estate (Auorbindo Ashram, Puducherry, India).

REFERENCES

1. Suthar, S. and Singh, S. Vermicomposting of domestic waste by using two epigeic earthworms (Perionyx excavatus and Perionyx sansibaricus). *International Journal of Environment Science and Technology*, **5**(1), 99–106 (2008).
2. Venkatesh, R. M. and Eevera, T. Mass reduction and recovery of nutrients through vermicomposting of fly ash. *Applied Ecology and Environmental Research*, **6**, 77–84 (2008).
3. Bansal, S. and Kapoor, K. K. Vermicomposting of crop residues and cattle dung with Eisenia foetida. *Bioresource Technology*, **73**(2), 95–98 (2000).
4. Singh, A. and Sharma S. Composting of a crop residue through treatment with microorganisms and subsequent vermicomposting. *Bioresource Technology*, **85**(2), 107–111 (2002).
5. Reddy, M. V. and Okhura, K. Vermicomposting of rice-straw and its effects on sorghum growth. *Tropical Ecology*, **45**, 327–331 (2004).
6. Lee, K. E. *Earthworms: Their Ecology and Relationships with Soils and Land Use*. Academic Press, London, UK (1985).
7. Atiyeh, R. M., Lee, S., Edwards, C. A., Arancon, N. Q., and Metzger, J. D. The influence of humic acids derived from earthworm-processed organic wastes on plant growth. *Bioresource Technology*, **84**(1), 7–14 (2002).
8. Kaushik, P. and Garg, V. K. Vermicomposting of mixed solid textile mill sludge and cow dung with the epigeic earthworm Eisenia foetida. *Bioresource Technology*, **90**(3), 311–316 (2003).
9. Kale, R. D., Bano, K., and Krishnamoorthy, R. V. Potential of Perionyx excavatus for utilizing organic wastes. *Pedobiologia*, **23**(6), 419–425.
10. Tomati, V., Grappel, A., Galli, E., and Rossi, W. Fertilizers from vermiculture—an option for organic wastes recovery. *Agrochimica*, **27**(2–3), 244–251 (1983).
11. Elvira, C., Sampedro, L., Benítez, E., and Nogales, R. Vermicomposting of sludges from paper mill and dairy industries with Eisena andrei: a pilot-scale study. *Bioresource Technology*, **63**(3), 205–211 (1998).
12. Garg, V. K. and Kaushik, P. Vermistabilization of textile mill sludge spiked with poultry droppings by an epigeic earthworm Eisenia foetida. *Bioresource Technology*, **96**(9), 1063–1071 (2005).
13. Suthar, S. Potential utilization of guar gum industrial waste in vermicompost production. *Bioresource Technology*, **97**(18), 2474–2477 (2006).
14. Graziano, P. L. and Casalicchio, G. Use of worm-casting techniques on sludges and municipal wastes: development and application. In: *On Eurthworms*, A. M. B. Pagliai and P. Omodeo (Eds.). Mucchi Editore, Modena, Italy, pp. 459–464 (1987).
15. Frederickson, J. and Knight, D. The use of anaerobically digested cattle solids for vermiculture. In: *Earthworms in Waste and Environmental Management*, C. A. Edwards and E. F. Neuhauser (Eds.). SPB Academie, Hague, Netherlands, pp. 33–47 (1988).
16. Frederickson, J., Butt, K. R., Morris, R. M., and Daniel, C. Combining vermiculture with traditional green waste composting systems. *Soil Biology and Biochemistry*, **29**(3–4), 725–730 (1997).
17. Karthikeyan, V., Sathyamoorthy, G. L., and Murugesan, R. Vermicomposting of market waste in Salem, Tamilnadu, India. In: *Proceedings of the International Conference on Sustainable Solid Waste Management*. Chennai, India, pp. 276–281 (September, 2007).

18. Jadia, C. D. and Fulekar, M. H. Vermicomposting of vegetable waste: a bio-physicochemical process based on hydro-operating bioreactor. *African Journal of Biotechnology*, **7**, 3723–3730 (2008).
19. Walkley, A. and Black, I. A. An examination of the Degtjareff method for determining soil organic matter and prepared modification of the chronic acid titration method. *Soil Science*, **34**, 29–38 (1934).
20. Jackson, M. L. *Soil Chemical Analysis*. Prentice Hall of India Private Limited, 1st edition, New Delhi, India, (1973).
21. Anderson, J. M. and Ingram, J. S. I. Soil organic matter and organic carbon. In: *Tropical Soil Biology and Fertility: A Hand Book of Methods*, J. M. Anderson and J. S. I. Ingram (Eds.). CAB International, Wallingford, UK, p. 221 (1993).
22. Simard, R. R. Ammonium acetate extractable elements. In: *Soil Sampling and Methods of Analysis*, R. Martin and S. Carter (Eds.). Lewis, Boca Raton, Florida, USA, pp. 39–43(1993).
23. Reinecke, A. J., Viljoen, S. A., and Saayman, R. J. The suitability of Eudrilus eugeniae, Perionyx excavates, and Eisenia fetida (Oligochaeta) for vermicomposting in southern Africa in terms of their temperature requirements. *Soil Biology and Biochemistry*, **24**, 1295–1307 (1992).
24. Edwards, C. A., Dominguez, J., and Neuhauser, E. F. Growth and reproduction of Perionyx excavatus (Perr.) (Megascolecidae) as factors in organic waste management. *Biology and Fertility of Soils*, **27**(2), 155–161 (1998).
25. Suthar, S. Nutrient changes and biodynamics of epigeic earthworm Perionyx excavatus (Perrier) during recycling of some agriculture wastes. *Bioresource Technology* 2007, **98**(8), 1608–1614.
26. Garg, P., Gupta, A., and Satya, S. Vermicomposting of different types of waste using Eisenia foetida a comparative study. *Bioresource Technology*, **97**(3), 391–395 (2006).
27. Nair, J., Sekiozoic, V., and Anda, M. Effect of pre-composting on vermicomposting of kitchen waste. *Bioresource Technology*, **97**(16), 2091–2095 (2006).
28. Wang, L. K., Nazih, K. S., and Yung-Tse, H. Vermicomposting process. In: *Biosolids Treatment Processes, vol. 6*, Humana Press, Totowa, New Jersey, USA, pp. 689–704 (2007).
29. Pramanik, P., Ghosh, G. K., Ghosal, P. K., and Banik, P. Changes in organic—C, N, P, and K and enzyme activities in vermicompost of biodegradable organic wastes under liming and microbial inoculants. *Bioresource Technology*, **98**(13), 2485–2494 (2007).
30. Gupta, R. and Garg, V. K. Vermiremediation and nutrient recovery of non-recyclable paper waste employing Eisenia fetida. *Journal of Hazardous Materials*, **162**(1), 430–439 (2009).
31. Kitturmath, M. S., Giraddi, R. S., and Basavaraj, B. Nutrient changes during earthworm, eudrilus eugeniae (Kinberg) mediated vermicomposting of agro-industrial wastes. *Karnataka Journal of Agriculture Science*, **20**, 653–654 (2007).
32. Edwards, C. A. and Dominguez, J. Vermicomposting of sewage sludge: effect of bulking materials on the growth and reproduction of the earthworm E. andrei. *Pedobiologia*, **44**(1), 24–32 (2000).
33. Peigne, J. and Girardin, P. Environmental impacts of farm—scale composting practices. *Water, Air, and Soil Pollution*, **153**(1–4), 45–68(2004).
34. Zibilske, L. M. Composting of organic wastes. In: *Principles and Applications of Soil Microbiology*, D. M. Sylvia, J. J. Fuhrmann, P. G. Hartel, and D. A. Zuberer (Eds.). Prentice Hall, Upper Saddle River, New Jersey, USA, pp. 482–497 (1999).
35. Nagavallemma, K. P., Wani, S. P., Stephane, L., et al. Vermicomposting: recycling wastes into valuable organic fertilizer. *Journal of SAT Agricultural Research*, **2**(1), 1–17 (2006).
36. Edwards, C. A. and Bater, J. E. The use of earthworms in environmental management. *Soil Biology and Biochemistry*, **24**(12), 1683–1689 (1992).
37. Singh, N. B., Khare, A. K., Bhargava, D. S., and Bhattacharya, S. Optimum moisture requirement during vermicomposting using Perionyx excavatus. *Applied Ecology and Environmental Research*, **2**, 53–62 (2004).
38. Liang, C., Das, K. C., and McClendon, R. W. The influence of temperature and moisture contents regimes on the aerobic microbial activity of a biosolids composting blend. *Bioresource Technology*, **86**(2), 131–137 (2003).

39. Mitchell, A. and Alter, D. Suppression of labile aluminium in acidic soils by the use of vermicompost extract. *Communications in Soil Science and Plant Analysis*, **24**(11–12), 1171–1181 (1993).
40. Haimi, J. and Huhta, V. Comparison of composts produced from identical wastes by vermistabilization and conventional composting. *Pedobiologia*, **30**(2), 137–144 (1987).
41. Tripathi, G. and Bhardwaj, P. Comparative studies on biomass production, life cycles and composting efficiency of Eisenia fetida (Savigny) and Lampito mauritii (Kinberg). *Bioresource Technology*, **92**(3), 275–283 (2004).
42. Loh, T. C., Lee, Y. C., Liang, J. B., and Tan, D. Vermicomposting of cattle and goat manures by Eisenia foetida and their growth and reproduction performance. *Bioresource Technology*, **96**(1), 111–114 (2005).
43. Ndegwa, P. M., Thompson, S. A., and Das, K. C. Effects of stocking density and feeding rate on vermicomposting of biosolids. *Bioresource Technology*, **71**(1), 5–12 (2000).
44. Guoxue, L., Zhang, F., Sun, Y., Wong, J. W. C., and Fang, M. Chemical evaluation of sewage composting as mature indicator for composting process. *Water Air Soil Sludge Pollution*, **132**, 333–345 (2001).
45. Crawford, J. H. Composting of agriculture waste. In: *Biotechnology: Applications and Research*, P. N. Cheremisinoff and R. P. Onellette (Ed.). Technomic Publishing, Lancaster, Pennsylvania, USA, 71 (1985).
46. Tognetti, C., Laos, F., Mazzarino, M. J., and Hernandez, M. T. Composting versus. vermicomposting: a comparison of end product quality. *Compost Science and Utilization*, **13**(1), 6–13 (2005).
47. Fares, F., Albalkhi, A., Dec, J., Bruns, M. A., and Bollag, J. M. Physicochemical characteristics of animal and municipal wastes decomposed in arid soils. *Journal of Environmental Quality*, **34**(4), 1392–1403 (2005).
48. Kaviraj and Sharma, S. Municipal solid waste management through vermicomposting employing exotic and local species of earthworms. *Bioresource Technology*, **90**(2), 169–173(2003).
49. Jadia, C. D. and Fulekar, M. H. Vermicomposting of vegetable waste: a bio-physicochemical process based on hydro-operating bioreactor. *African Journal of Biotechnology*, **7**, 3723–3730 (2008).
50. Edwards, C. A. *Earthworm Ecology* 2nd ed., CRC Press LLC, Boca Raton, Florida, USA (2004).
51. Albanell, E., Plaixats, J., and Carbrero, T. Chemical changes during vermicomposting (Eisenia fetida) of sheep manure mixed with cotton industrial wastes. *Biology and Fertility of Soils*, **6**, 266–269 (1988).
52. Tognetti, C., Mazzarino, M. J., and Laos, F. Improving the quality of municipal organic waste compost. *Bioresource Technology*, **9**, 1067–1076 (2007).
53. Short, J. C. P., Frederickson, J., and Morris, R. M. Evaluation of traditional windrow-composting and vermicomposting for the stabilization of waste paper sludge (WPS). *Pedobiologia*, **43**, 735–743 (1999).
54. Saradha, T. The culture of earthworms in the mixture of pond soil and leaf litter and analysis of vermi fertilizer. *Journal of Ecobiology*, **9**, 185–188 (1997).
55. Uthaiah, P. A. *Acceleration of pressmud decomposition by microbial inoculation for quality product*. M. S. thesis. University of Agricultural Sciences, Bangalore, India (1997).
56. Muthukumarasamy, R., Revathi, G., Murthy, V., Mala, S. R., Vedivelu, M., and Solayappan, A. R. An alternative carrier material for bio-fertilizers. *Co-Operative Sugar*, **28**, 677–680 (1997).
57. Parthasarathi, K. and Ranganathan, L. S. Aging effect on enzyme activities in pressmud vermicasts of Lampito mauritii (Kinberg) and Eudrilus eugeniae (Kinberg). *Biology and Fertility of Soils*, **30**(4), 347–350 (2000).
58. Khwairakpam, M. and Bhargava, R. Vermitechnology for sewage sludge recycling. *Journal of Hazardous Materials*, **161**(2–3), 948–954 (2009).
59. Senapati, B. K. Vermitechnology: an option for recycling cellulosic waste in India. In: *New Trends in Biotechnology*. Oxford and IBH Publications, Calcutta, India, pp. 347–358 (1992).

60. Padmavathiamma, P. K., Li, L. Y., and Kumari, U. R. An experimental study of vermi-biowaste composting for agricultural soil improvement. *Bioresource Technology*, **99**(6), 1672–1681 (2008).
61. Sangwan, P., Kaushik, C. P., and Garg, V. K. Vermiconversion of industrial sludge for recycling the nutrients. *Bioresource Technology*, **99**(18), 8699–8704 (2008).
62. Bouche, M., Al-addan, F., Cortez, J., et al. Role of earthworms in the N cycle: a falsifiable assessment. *Soil Biology and Biochemistry*, **29**(3–4), 375–380 (1997).
63. Balamurugan, V., Gobi, M., and Vijayalakshmi, G. Comparative studies on degradation of press mud using cellulolytic fungi and exotic species of earthworms with a note on its gut microflora. *Asian Journal of Microbiology, Biotechnology and Environmental Sciences*, **1**, 131–134 (1999).
64. Atiyeh, R. M., Lee, S., Edwards, C. A., Subler, S., and Metzger, J. D. Earthworm processed organic wastes as components of horticulture potting media for growing marigolds and vegetable seedlings. *Compost Science and Utilization*, **8**, 215–223 (2000).
65. Crawford, J. H. Review of composting. *Process of Biochemistry*, **8**, 14–15 (1983).
66. Cabrera, M. L., Kissel, D. E., and Vigil, M. F. Nitrogen mineralization from organic residues: research opportunities. *Journal of Environmental Quality*, **34**(1), 75–79 (2005).
67. Fang, M., Wong, M. H., and Wong, J. W. C. Digestion activity of thermophilic bacteria isolated from ash-amended sewage sludge compost. *Water Air and Soil Pollution*, **126**, 1–12 (2001).
68. Morais, F. M. C. and Queda, C. A. C. Study of storage influence on evolution of stability and maturity properties of MSW composts. In: Proceedings of the 4th International Conference of ORBIT Association on Biological Processing of Organics: *Advances for a Sustainable Society*, Perth, Australia (May, 2003).
69. Haug, R. T. The Practical Handbook of Compost Engineering, 2nd ed. Lewis, CRC Press, Boca Raton, Florida, USA (1993).
70. Dash, M. C. and Senapati, B. K. Vermitechnology, an option for organic wastes management in India. In: *Verms and Vermicomposting*. M. C. Dash, B. K. Senapati, and P. C. Mishra (Eds.). Sambalpur University, Sambalpur, Orissa, India, pp. 157–172 (1986).
71. Talashilkar, S. C., Bhangarath, P. P., and Mehta, V. B. Changes in chemical properties during composting of organic residues as influenced by earthworm activity. *Journal of the Indian Society of Soil Science*, **47**, 50–53 (1999).
72. Christry, M. A. V. and Ramaligam, R. Vermicomposting of sago industrial soild waste using epigeic earthworm Eudrilus eugeniae and macronutrients analysis of vermicompost. *Asian Journal of Microbiology, Biotechnology and Environmental Sciences*, **7**, 377–381 (2005).
73. Manna, M. C., Jha, S., Ghosh, P. K., and Acharya, C. L. Comparative efficacy of three epigeic earthworms under different deciduous forest litters decomposition. *Bioresource Technology*, **88**(3), 197–206 (2003).
74. Ghosh, M., Chattopadhyay, G. N., and Baral, K. Transformation of phosphorus during vermicomposting. *Bioresource Technology*, **69**, 149–154 (1999).
75. Delgado, M., Bigeriego, M., Walter, I., and Calbo, R. Use of California red worm in sewage sludge transformation. *Turrialba*, **45**, 33–41 (1995).
76. Rao, S., Rao, A. S., and Takkar, P. N. Changes in different forms of K under earthworm activity. In: *Proceedings of the National Seminar on Organic Farming and Sustainable Agriculture*, Ghaziabad, India, pp. 9–11 (October, 1996).
77. Tiwari, S. C., Tiwari, B. K., and Mishra, R. R. Microbial populations, enzyme activities and nitrogen–phosphorous–potassium enrichment in earthworm casts and in the surrounding soil of a pineapple plantation. *Biology and Fertility of Soils*, **8**, 111 (1989).

9 Earthworms and Nitrous Oxide Emissions

Andrew K. Evers, Tyler A. Demers, Andrew M. Gordon, and Naresh V. Thevathasan

CONTENTS

9.1 INTRODUCTION

The presence of earthworms can be seen as an added benefit to many agricultural systems since earthworms contribute greatly to the overall physical properties of agricultural soils [1]. The study in sole cropping systems have focused on the ability of earthworms to facilitate soil mixing and the decomposition of organic matter which is especially important in agricultural systems [2-4]. Earthworms also affect soil properties by increasing soil porosity and decreasing bulk density through bioturbation and cast deposition on the soil surface [1]. Earthworm activity stimulates mineralization of N in residues which promotes the availability for plants and microorganisms of inorganic forms of N from plant material [1, 5].

However, increased earthworm population might increase the production of nitrous oxide (N_2O) emissions from agricultural soils. Over 50% of *in situ* N_2O emissions, in some soils, could be a result of earthworm activity [6]. Recent research suggests that, globally, earthworms could be producing up to 3×10^8 kg of N_2O annually [6]. Conventional agricultural practices which aim to encourage earthworm populations due to their positive influence on soil properties are the highest anthropogenic sources of N_2O emissions. On a global scale, annual emissions of N_2O were 16.2 Tg in 2004 [7], and as a result, earthworms could be responsible for nearly 2% of global emissions.

One reason for this is that the earthworm gut is an ideal environment for denitrification [8-10]. Using microsensors, Horn et al. [9] determined that the earthworm gut is anoxic and contains copious carbon substrates for microorganisms and is therefore ideal for N_2O production. Denitrification is enhanced when the earthworm ingests denitrifier bacteria with organic matter [1, 8-10]. When gaseous N_2O is produced, it is able to escape the permeable epidermis of the earthworm and diffuses from the soil surface [9].

At the Guelph Agroforestry Research Station (GARS) in Guelph, Ontario, Canada, Price and Gordon [11] found that earthworm density was greater in a tree-based intercropping (TBI) system than in a conventional agricultural monoculture. A TBI system is defined as "an approach to land use that incorporates trees into farming systems and allows for the production of trees and crops or livestock from the same piece of land in order to obtain economic, ecological, environmental, and cultural benefits" [12]. These systems incorporate leaf litter and increase soil water content which could encourage higher earthworm populations compared to sole cropping systems. In turn, this could increase the overall volume of the earthworm gut, thereby facilitating denitrification and higher N_2O emissions from a TBI system. Price and Gordon [11] also speculated that the reason earthworm densities were higher in the intercropped system compared to the conventional monoculture was because earthworms move to an area with a lower soil temperature, which in turn are areas that also have higher soil water content.

Currently, very little information exists on the influence that earthworm density has on N_2O emissions from agricultural soils, and specifically those potentially associated with a TBI system. The objective of this study was to investigate the relationship, if any, between N_2O flux, earthworm density, and gravimetric soil water content, taking into account the earthworm densities calculated by Price [13] in the TBI and monoculture systems located at GARS and using the most common earthworm species found in GARS, the common nightcrawler (Lumbricus terrestris L). It was hypothesized that N_2O flux would be higher as earthworm density and soil water content increased.

9.2 MATERIALS AND METHODS

9.2.1 Study Design

The first experiment was conducted in the Science Complex Phytotron at the University of Guelph, Ontario, and Canada. The purpose of the first experiment was to determine the optimal soil water content for earthworm activity resulting in the highest

N_2O emissions. The experiment was a two factorial, completely random design with four replications for a total of 64 experimental units. The first factor was earthworm density and the second factor was gravimetric soil water content (15, 25, 35, and 45%).

Soil was collected from GARS and homogenized using a 2 mm sieve. The soil is sandy loam in texture with an average pH of 7.2 [14]. A leaf litter mixture composed of silver maple (*Acer saccharinum L.*) and poplar (*Populus spp.)* leaves was also collected from GARS, dried at 60°C for 1 week, and mixed into the homogenized soil to achieve a soil organic matter content of approximately 3%. The 4 kg of the soil and leaf litter mixture was then put into each of the 5 L polypropylene mesocosms, equipped with an airtight lid and rubber septum for sampling. The lids were only placed on the mesocosms at the time of N_2O sampling. The surface area of each mesocosm was 0.033 m^2.

Earthworm density was calculated based on data collected in the spring of 1997 from GARS by Price [13]. The three earthworm densities included high, medium, and low earthworm densities, representing populations found 0 , 3 , and 6 m from the tree row in a TBI system, respectively. However, these values were tripled in order to ensure the detection of N_2O for the purpose of finding optimal soil water content and also to represent an earthworm invasion where populations could initially be very high and decline over time [15]. These values were 30, 20, and 10 earthworms per 4 kg of soil or 0.033 m, for the high, medium, and low treatments, respectively, and a control with no earthworms. *L. terrestris* were purchased from Kingsway Sports (Guelph, Ontario, Canada). Earthworms were counted and weighed prior to being added to the mesocosms.

Prior to adding the earthworms, each mesocosm was fertilized with urea (46-0-0, N-P-K), which represented the N fertilizer requirement for corn planted at GARS (215 kg ha^{-1}). Deionized water was applied to each mesocosm for 1 week prior to adding the earthworms in order to achieve the desired gravimetric soil water content for each treatment. A small hole in the bottom of each mesocosm allowed for proper drainage. During the course of the experiment, soil water content was maintained by weight. The mesocosms were weighed every day for the entire course of the experiment and deionized water was added to bring each mesocosm to the desired water content.

The mesocosms were placed in a greenhouse with a constant air temperature of 20°C and monitored light conditions of 16/8 hr cycles. Soil temperature was monitored using Priva soil temperature sensors (Priva North America Inc., Vineland Station, Ontario, Canada) to ensure a constant soil temperature of approximately 20°C. The N_2O sampling technique and calculations will be explained in this study.

A second experiment was conducted from February 2009 to March 2009 in the Science Complex Phytotron at the University of Guelph. Experiment 2 was a completely random design with four replications for a total of 16 experimental units. A control with no earthworms and earthworm densities of 9 (high), 6 (medium), and 3 (low) earthworms per mesocosm were used for a total of four treatments. The high, medium, and low density treatments were calculated based on actual densities found by Price [13] at GARS representing an earthworm density adjacent to the tree row, 3 m from the tree row, and 6 m from the tree row in a TBI system, respectively; a control with no earthworms was also included.

Optimal gravimetric soil water content was determined in Experiment 1 and was found to be 31%. This soil water content treatment was held constant for all four earthworm density treatments over the duration of the experiment. Methods for soil preparation, maintaining gravimetric soil water content, and monitoring temperature were the same as in Experiment 1.

9.2.2 Sampling Procedure

At the time of N_2O sampling, the airtight lid was placed onto each mesocosm and a 30 mL air sample using a 26 gauge needle and syringe was taken at t = 0, 30, and 60 min to calculate N_2O flux over an hour. Air samples were deposited into 12 mL Exetainers (Labco Limited, United Kingdom) and analyzed using a SRI Model 8610C Gas Chromatograph (Torrance, California, USA) at environment Canada (Burlington, Ontario, Canada). The N_2O samples were taken once a week for 4 weeks beginning at 10:00 AM.

A soil sample was taken from each mesocosm, both before the addition of earthworms and after the last week of sampling. This was done to measure the initial and the final nitrate (NO_3^-), ammonium (NH_4^+), and total inorganic N (TIN) concentrations to determine if there was a change over the course of the experiment. Soil samples were stored in the freezer until analysis. N content was measured following a 2N KCl extraction [16], and samples were run through an Astoria 2-311 Analyzer (Astoria-Pacific Inc., Oregon, USA). Measurements of soil inorganic and organic carbon (C) were also done for initial and final C content using a Leco C determinator (Leco Corporation, St Joseph, MI, U.S.A.). However, results for soil N and C are not reported here and are part of a larger study.

9.2.3 N_2O Flux Calculation

The N_2O flux was calculated using the ideal gas law; the molar volume of N_2O at 0°C and 1 atm is 44.0128 L/mol. The N_2O flux was adjusted for air temperature and pressure using the following formula:

$$\text{Flux adjustment} = \begin{array}{l} 44.0128\ \text{Lmol}^{-1} * \dfrac{[(23.16°K + T°C)]}{273.16°K} \\ * \dfrac{(1013.2hPa)}{PhPa} \end{array} \qquad (1)$$

where T is the air temperature and P is the air pressure on the day of sampling. These values were taken into consideration because a temperature greater than 0°C increases molar volume and, air pressure that is greater than atmospheric decreases molar volume.

The volume of the mesocosm was then converted to mole of air and multiplied by the slope of the flux determined by hourly measurement. This value was then used to calculate the flux in μmol m^{-2} s^{-1}:

$$\text{Flux}\left(\mu\text{mol m}^{-2}s^{-1}\right)=\frac{\left(S\ nmol\ mol^{-1}s^{-1}\right)\left(M\ mol\right)}{Xm^2} \tag{2}$$

where S represents the slope of the line (N_2O concentration at each measurement interval over 1 hr), M is the molar volume of the air in the mesocosm, and X represents the area of the mesocosm. This value was then converted into kg of N_2O ha^{-1} day^{-1}:

$$\begin{aligned}\text{Flux}\left(g\ ha^{-1}day^{(-1)}\right)&=\left(F\mu mol\ m^{-2}s^{-1}\right)\left(1.0\times10^{-9}mol\right)\\&\times\left(44.0128\ L\ mol^{-1}\right)\left(10000m^2\right)\\&\times\left(86400s\right)\left(1000g\right)\end{aligned} \tag{3}$$

where F is the flux calculated from (2).

Some of the flux values were negative as a result of a sink of N_2O being created rather than the N_2O being emitted through the soil surface during the extraction period from the mesocosms which created negative flux values [17]. Therefore, a value of 100 was added to all flux values in order to complete statistical analyses and maintain positive values since the statistical program could only read positive values. The final flux values following analysis were then subtracted by hundred to present actual flux values in the following statistical analysis.

9.2.4 Statistical Analysis

All statistical tests were conducted using SAS v.9.1 (SAS Institute, Cary, NC, USA) at an error rate of $\alpha = 0.05$. An analysis of variance (ANOVA) using repeated measures in the PROC MIXED function was used to compare the effects of earthworm density and N_2O flux according to soil water content treatment to determine the variance in initial and final earthworm biomass between moisture treatments, as well as mortality rates between moisture treatments in Experiments 1 and 2. A response surface design using the PROC RSREG function [18] was applied to data from Experiment 1 to determine the optimal range levels of earthworm density and soil water content for the production of N_2O over ranges for these parameters that were not part of the original experimental design. The optimal soil water content found through the RSREG was then applied to Experiment 2.

9.3 DISCUSSION

Overall, emissions were highest at the 25 and 35% soil water content treatments and the lowest emissions were seen at 15 and 45% soil water content. Bertora et al. [19] found similar results with the presence of earthworms, where emissions increased significantly over the course of their experiment at 25% soil water content up to 62 days, when emissions began to decrease. The N_2O emissions were significantly higher at 25% than at the lower moisture treatments (19 and 12.5%) where emissions were not significant.

Conversely, at 35% moisture, there was a downward trend in emissions over the course of the experiment, except at the high density where N_2O flux peaked at 69.6 g ha^{-1} day^{-1} in 2nd week with a significant decline in emissions in 3rd week. This could mean that earthworms may only be able to tolerate high soil water content for a limited time. Therefore, the high earthworm mortality in this treatment could have occurred toward the end of the experiment, which could explain the decline in N_2O flux following 2nd week. However, the 45% soil water content treatment also contradicts optimal soil water content for earthworm activity. The N_2O emissions gradually increased across all earthworm densities at 45% soil water content showing that L. terrestris may have been adapting to the soil conditions. Increases in emissions were gradual and did not reach levels found at 25 and 35% soil water content, but mortality rates were lower, but not significant, compared to mortality rates at 35% soil water content showing some tolerance. El-Duweini and Ghabbour [20] also reported soil water content tolerance levels but for two Australian species, *Allolobophora caliginosa* and *Metaphire californica*, to be 20–45% and 35–55%, respectively in a clay soil.

Earthworm mortality was the highest in the 35% moisture treatment, at the highest earthworm density even though emissions were significantly higher than at any other treatment combination. Dymond et al. [15] reported an initial earthworm invasion of 2,621 individual's m^{-2} of Dendrobaena octaedra into a northern Alberta pine (*Populus sp.*) and aspen (*pinus sp.)* forest. This population dropped to 76 individuals m^{-2} within just a few years as a result of competition for resources. High competition could have been the reason for the drastic decline in emissions in the high density treatment at 35% soil water content and lower mortality in the medium (9%) and low (0%). Another reason for the decline in emissions after 2 week could be due to the ability of high earthworm populations to speed up residue decomposition [19]. Organic matter is more palatable to earthworms at higher soil water content; therefore, ingestion of organic matter is enhanced. Organic matter turnover could have been enhanced at the 35% moisture and high density combination by 2 week resulting in a decrease in pre-ingested organic matter and a decline in earthworm activity.

The gravimetric soil water content treatments of 15, 25, 35, and 45% are approximately equivalent to a water-filled pore space (WFPS) of 30, 55, 75, and 100%. It is generally accepted that denitrification rates are optimal between a WFPS between 60 and 100%, where N_2O is the primary product between 60 and 90% [21]. Above 90% N is the dominant product [21] which could be the reason for lower N_2O flux measurements in the 45% soil water content, where the WFPS was 100%. The highest N_2O flux occurred at 35% soil water content or 75% WFPS, which is within the range of optimal denitirification rates. Furthermore, nitrification rates are highest between 45 and 75% [21]. The product of nitrification is NO_3^-, a primary input for denitrification. This means that in the 35% soil water content treatment, both nitrification and denitrification were optimal, which may have contributed to the significantly higher N_2O flux compared to the 15 and 45% soil water content treatments.

The N_2O emissions could be lower at dryer soil water contents as a result of earthworm diapause or aestivation. In this state, earthworms will decrease their activity to prevent water loss from the body [2]. Ingestion of soil and organic matter content would decrease, thereby limiting microbial activity in the earthworm gut and reducing emissions. The same occurs at high moisture contents and could explain the lower N_2O emissions at the 45% moisture treatment in this study.

Perrault and Whalen [22] found that earthworm burrowing length decreased in wetter soils, which would indicate a decrease in earthworm activity. However, wetter soils caused an increase in the ingestion of organic matter compared to dryer soils. Leaf litter is more palatable to earthworms when wetted, and as a result ingestion is increased. This could explain higher emissions in the 25 and 35% moisture treatments compared to 15% soil water content, as well as the decrease in earthworm biomass in the medium and high densities at 15% soil water content. Earthworms would ingest higher carbon substrates at these moisture contents which would in turn provide energy for denitrifying bacteria found in the earthworm gut and increase N_2O production.

Earthworm surface casting also increases in wetter soils which provide another ideal environment for denitrification. Earthworm casts contain higher populations of denitrifying bacteria compared to mineral soils due to higher amounts of carbon substrates, and as a result, higher N_2O emissions are produced [23]. Elliot et al. [24] found that denitrification was higher in earthworm casts than surrounding mineral soil. Denitrification rates from earthworm casts ranged between 0.2–0.9 $g\,N\,g^{-1}$ during the fall compared to 0.05–0.3 $g\,N\,g^{-1}$ from the soil within the same time period. This indicates that a portion of the emissions from this experiment could be due to increased surface casting in the 25 and 35% moisture treatments at the high density treatments.

Trends in N_2O emissions according to earthworm density did occur. The high, medium, and low density treatments represent 9.1×10^5, 6.1×10^5, and 3.0×10^5 earthworms ha^{-1}, respectively. Emissions consistently increased as earthworm density increased in all moisture treatments. However, emissions were only significantly higher at the medium and high densities in the 25% and 35% soil water content treatments (Figure 1). Frederickson and Howell [25] found no relationship between earthworm density and N_2O emissions in large-scale vermicomposting beds. However, in a subsequent laboratory experiment, emissions were correlated with earthworm density at five earthworm density treatments ($R^2 = 0.76$).

The reason for this may be a result of an increase in the ingestion of organic matter and, with that, denitrifier bacteria at higher earthworm densities; therefore, denitrification may occur at faster rates than in soils with lower earthworm densities. Denitrification occurs at higher rates in the earthworm gut due to the anoxic environment and sufficient supply of carbon for denitrifier bacteria compared to soil homogenates [6, 26, 27]. An increase in earthworm density results in an increase in this ideal environment of earthworms for denitrifier bacteria and therefore, could increase emissions. The number of denitrifier bacteria is also higher in the earthworm gut and surface castings than outside soil homogenates [8]. These are calculated that there were 256-fold more denitrifier bacteria in the earthworm gut of *L. rubellus* than in the surrounding soil where the earthworms were found. This indicates that an increase in earthworm density also increases the number of denitrifier bacteria in the gut of the earthworms facilitating higher N_2O emissions as could be the case in this study.

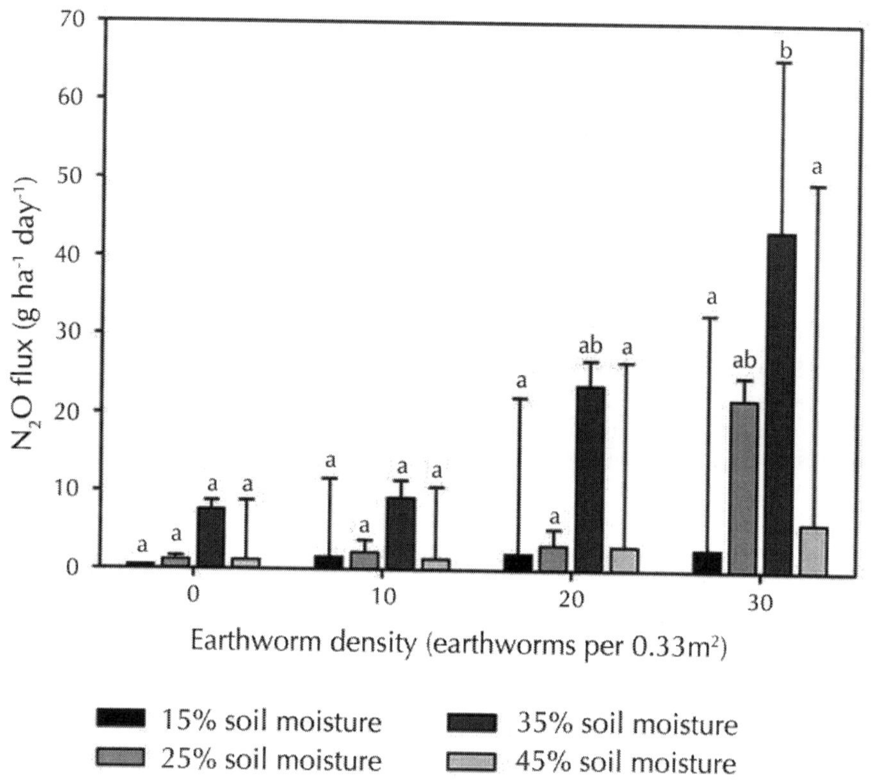

FIGURE 1 The relationship between N_2O emissions, earthworm density per 0.033 m^2, and gravimetric soil water content (P = 0.057, SE = 4.07, 5.29, 6.10, and 4.52 for 15, 25, 35, and 45%, resp.). Bars with same letter indicate no significant difference between treatments at P = 0.05 according to Tukey-Kramer means adjustment.

Another reason why N_2O emissions were highest at the high density earthworm treatments could have been a result of an increase in the microbial biomass pool and subsequent increase in respiration causing lower O_2 levels in the soil. Groffman et al. [28] found that in areas with the presence of earthworms, microbial biomass was significantly higher in the mineral soil compared to areas without the presence of earthworms. In turn, Fisk et al. [29] discovered that this increase in microbial biomass due to the presence of earthworms increased respiration rates by 20% compared to areas without earthworms. Therefore, O_2 levels will decline providing a more ideal environment for denitrification to occur and subsequent gaseous N losses. However, even though microbial biomass may increase with earthworm presence, a subsequent increase in mineralization and nitrification rates may not occur. Bohlen et al. [30] and Groffman et al. [28] found that mineralization and nitrification rates in the soil did not differ significantly in plots with and without earthworms. They speculated that earthworms facilitated a C-sink in the soil and subsequently created an N-sink, preventing the increase in N mineralization and nitrification rates in the soil. This could mean that the majority of the N_2O released from the mesocosms was attributed to the presence of earthworms and earthworm gut, rather than denitrification occurring in the surrounding soil, since NO_3^- concentrations may have been low due to low nitrification rates.

In Experiment 2, N_2O emissions were not significantly different across earthworm population densities; however, the results were consistent to what was found in Experiment 1. N_2O flux in Experiment across all earthworm densities (0, 3, 6, and 9) was in the same range as emissions in Experiment 1 between the control and low density earthworm treatments (0 and 10). This was expected since the earthworm densities used in Experiment 2 were within the range of the control and low density treatments in Experiment 1, and there were no significant differences in emissions between the control and low density treatments in Experiment 1. No significant differences in N_2O flux occurred even with significant differences in initial and final biomass between density treatments. This shows that even with an increase of approximately 3×10^5 earthworms ha^{-1} from zero earthworms, there would be no significant corresponding change in emissions between a TBI and sole cropping system, like the systems found at GARS. This could be a result of other compounding factors such as soil water content, soil temperature, residual soil N and C, and land management practices, which could all mask the earthworm effect on denitrification.

The same general trend of N_2O emissions occurred over time as in the 35% soil water content treatment in Experiment 1, where emissions hit a peak at week 2 and declined at week 3 to levels the same or lower than at week 1. This cannot be explained by earthworm mortality since earthworm mortality was insignificant or did not occur in Experiment 2 compared to Experiment 1. However, since soil water content of 31% was found to be optimal for earthworm activity, this may have sped up organic matter decomposition [18] between weeks 1 and 2 leaving the less palatable lignin compounds, thereby slowing earthworm activity between weeks 2 and 3.

9.4 RESULTS

9.4.1 N_2O Emissions

The earthworm density and soil water content interaction on N_2O emissions was significant ($P = 0.0457$). Mean N_2O emissions ranged from 0.54 g ha^{-1} day^{-1} from the 15% moisture and no earthworm density treatment to 43.5 g ha^{-1} day^{-1} from the 35% moisture and high earthworm density treatment as illustrated in Figure 1. Patterns did exist in emissions, where N_2O emissions were highest at the high density across all moisture treatments and lowest in the mesocosms with no earthworms across all moisture treatments. The extent of emissions across all of the moisture treatments was high > medium > low > control. Emissions due to moisture were 35% > 25% > 45% > 15% across all earthworm densities except when earthworm density = 0, where emissions were 35% > 25% > 45% =15%. Emissions were only significant at the high density and 25 and 35% soil water content treatments, as well as the medium density and 35% soil water content treatment compared to the rest of the treatments.

Over the course of the experiment, N_2O emissions only increased at 45% soil water content, where emissions were highest in the last week of sampling compared to the 1st week at all density treatments (Figure 2). At 15 and 25% soil water content, emissions peaked at week 3 and 2, respectively, and declined by week 4. In the 25% moisture treatment, emissions had a significant peak at 56.6 g ha^{-1} day^{-1} at high density in week 2 compared to 1.5, 3.2, and 3.6 g ha^{-1} day^{-1} in the control, low, and medium densities, respectively. An outlier did exist in the 25% moisture and high density treatment during week 2, but when left in, it did not significantly change the result. However, it may explain the peak in emissions during week 2 at the high density treatment. N_2O emissions declined over the course of the experiment in all densities at 35% moisture except in the high density treatment where emissions were the highest in week 2 at 69.6 g ha^{-1} day^{-1}.

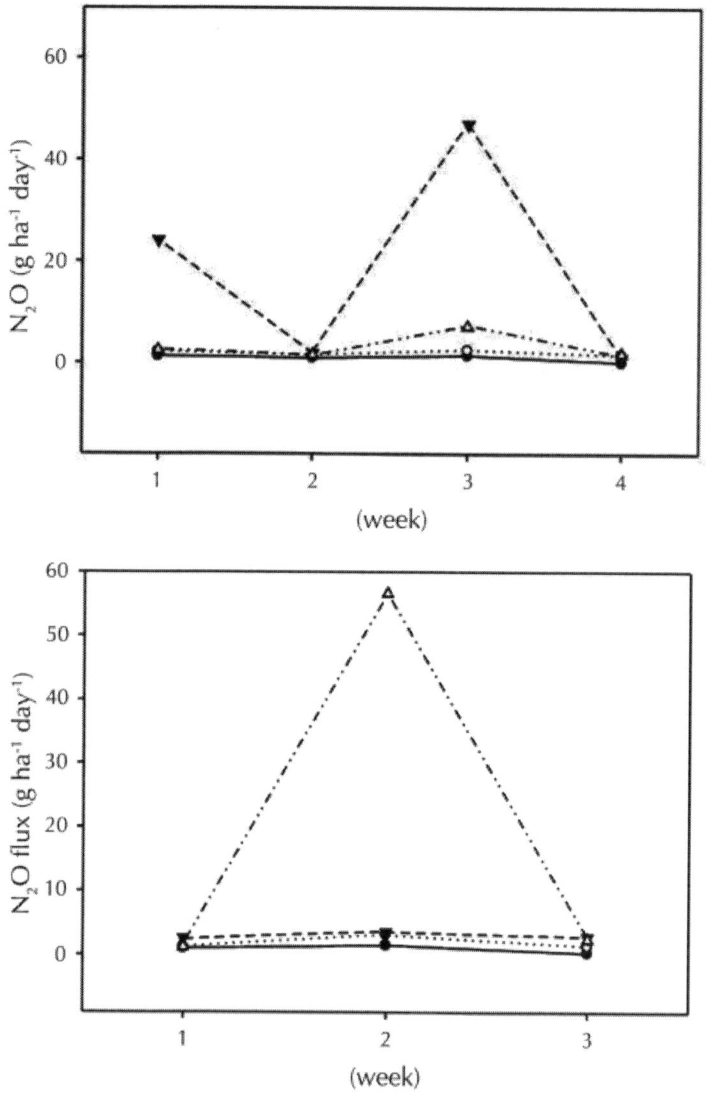

FIGURE 2 *(Continued)*

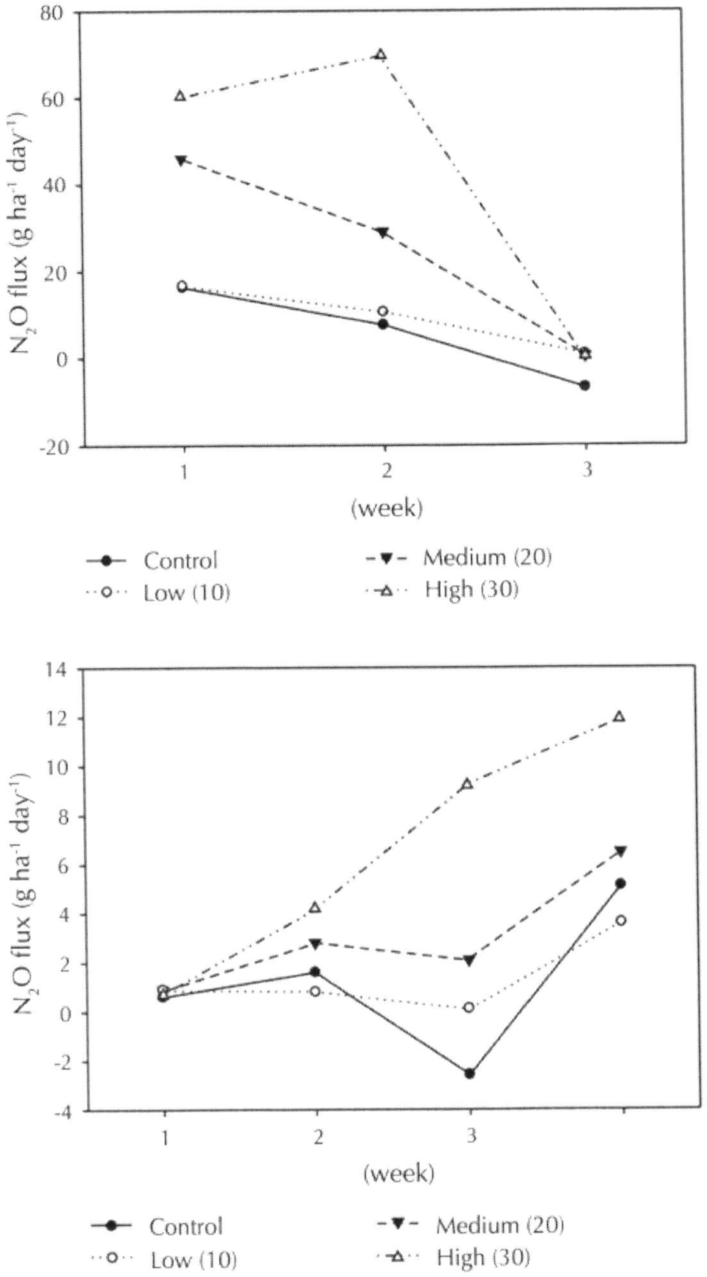

FIGURE 2 The N_2O flux over the entire course of the experiment according to the control, low, medium, and high earthworm density at (a) 15% ($P = 0.1398$), (b) 25% ($P = 0.3912$), (c) 35% ($P = 0.2451$), and (d) 45% ($P = 0.0685$) gravimetric soil water content.

A response surface regression indicated that the lowest N_2O emissions would occur at soil water content of 15% and an earthworm density of 13 earthworms 0.033 m^{-2}, whereas the highest emissions would occur at a soil water content of approximately 31% and an earthworm density of 30 individuals 0.033 m^{-2} as seen in Figure 3. The lowest and highest emissions correspond to −1.7 and 22.3 g ha^{-1} day^{-1}, respectively. These numbers represent emissions within the treatment range of the experiment. Emissions at soil water content or earthworm density outside of the treatment range can be determined using the equation found in the caption for Figure 3.

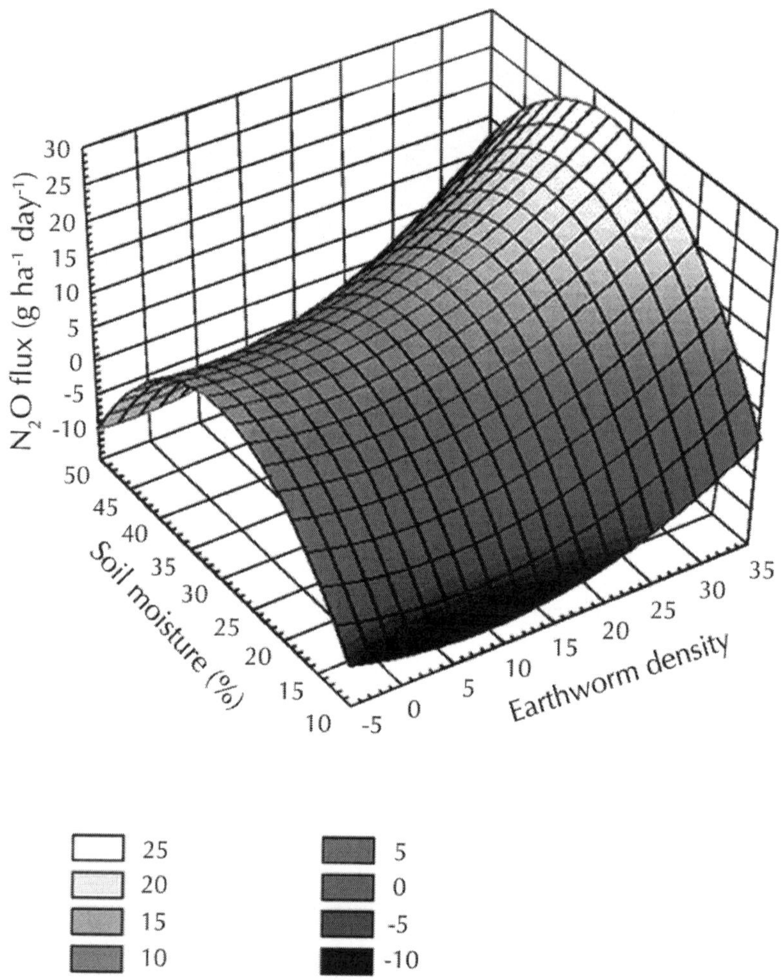

FIGURE 3 A response surface regression showing the relationship between N_2O flux (kg ha^{-1} day^{-1}), gravimetric soil water content (%), and earthworm density (number of earthworms 0.033 m^{-2}). Equation of the line is $36.7186 - (0.36143 \times D) + (3.1095 \times M) + (0.0174 \times D \times D) + (0.00810 \times M \times D) - (0.0518 \times M \times M)$ ($R^2 = 0.17$, $P \leq 0.0001$), where is earthworm density and is gravimetric soil water content.

9.4.2 Earthworm Mortality and Biomass

Mortality rates were not significantly different between moisture treatments within the density treatments (Table 1). There was very little mortality in the low-density treatment across all soil moisture treatments. Mean mortality rates in the medium density treatment ranged from 3 to 11%, the highest mortality rate occurring in the 15% moisture treatment and the lowest in the 25% moisture treatment. Mean mortality in the high-density treatment ranged from 5 to 18%, the highest mortality rate occurring in the 35% soil moisture treatment and the lowest occurring in the 25% moisture treatment.

TABLE 1 Mean earthworm mortality in the low, medium, and high earthworm densities according to θ_g (%) treatment.

	Mortality Rate (%)		
θ_g (%)	Low§	Medium	High
15	2.0 a†	11.0 a	15.6 a
25	2.0 a	2.5 a	5.0 a
35	0.0 a	8.5 a	18.3 a
45	0.0 a	6.0 a	10.0 a
SE	0.6	0.6	1.0
value	.1994	.2139	.0571

†Within columns, means followed by the same letter are not significantly different according to Tukey-Kramer means adjustment (0.05).

§Low, medium, and high refer to densities of earthworms per 0.033 m^2: 10, 20, and 30, respectively.

The difference in the initial and final earthworm biomass was significant according to soil water content across all earthworm density treatments as seen in Table 2. The largest increase in biomass in the low density treatment also occurred at 35% soil water content. The largest increase in biomass in the medium density treatment occurred at 35% soil water content where the final earthworm biomass was significantly higher than the initial biomass. Earthworm biomass declined in the 15% soil water content treatment due to a mortality rate of 11%; however, this decline was not significant. The highest increase in earthworm biomass over the course of the experiment occurred at 25% soil water content in the high density treatment; however, this increase was not significant. There was also a decline in earthworm biomass over the course of the experiment in the 15 and 35% soil water content treatment due to high mortality rates in the high density treatment; however this decline was not significant.

TABLE 2 Mean initial and final earthworm biomass in the low, medium and high earthworm densities according to θ_g (%) treatment.

	Density Treatment Biomass (g)		
θ_g (%)	Low§	Medium	High
15 Initial	40.88 a†	86.97 a	130.55 abc
15 Final	50.92 a	83.85 a	121.48 abc
25 Initial	43.08 a	86.97 a	125.85 abc
25 Final	52.19 a	101.95 b	150.85 c
35 Initial	47.20 a	103.60 b	145.63 ac
35 Final	89.94 b	123.30 c	144.50 ac
45 Initial	33.28 c	70.29 a	103.90 b
45 Final	55.59 a	75.60 a	117.10 ab
SE	2.69	4.02	5.54
value	<.0001	.0420	.0323

†Within columns, means followed by the same letter are not significantly different according to Tukey-Kramer means adjustment (0.05).
§Low, medium, and high refer to densities of earthworms per 0.033 m^2: 10, 20, and 30, respectively.

9.4.3 N_2O Emission at 31% Gravimetric Soil Water Content

Based on the gravimetric soil water content of 31% found in the response surface in Experiment 1, there was no significant difference in N_2O flux across all earthworm densities ($P = 0.8085$). Mean N_2O flux over the duration of the experiment was 6.99, 5.49, 6.36, and 5.63 g ha^{-1} day^{-1} for the control, low, medium, and high earthworm densities, respectively. There was also no significant difference in mean N_2O flux according to the week by density interaction ($P = 0.7611$, SE = 2.37 for the control,SE = 2.05 for low, medium, and high earthworm density). However, at all earthworm densities, N_2O flux peaked at week two and then declined below week one levels at week 3.

9.4.4 Earthworm Mortality and Biomass at 31% Soil Water Content

Earthworm survival was 100% in the low and medium density treatments and 95% in the high density treatment. Initial and final earthworm biomass was significantly different across all earthworm densities. Earthworm biomass in the low density treatment increased from 12.6 g at the start of the experiment to 27.1 g at the end ($P = 0.0007$) as seen in Table 3. In the medium density treatment the initial earthworm biomass was

28.6 g and increased to 44.4 g by the end of the experiment (P = 0.0003). In the high density treatment, earthworm biomass increased from 36.9 to 74.2 g by the completion of the experiment (P = 0.0001).

TABLE 3 Mean initial and final biomass in the low, medium, and high earthworm densities at 31% gravimetric soil moisture content.

	Density Treatment Biomass (g)		
	Low§	**Medium**	**High**
Initial	12.6 a†	28.6 a	36.9 a
Final	27.1 b	44.4 b	74.2 b
SE	1.94	1.94	1.94
P value	.0007	.0003	.0001

†Within columns, means followed by the same letter are not significantly different according to Tukey-Kramer means adjustment (0.05).
§Low, medium, and high refer to densities of earthworms per 0.033 m^2: 3, 6, and 9, respectively.

9.5 CONCLUSION

A relationship was found between earthworm density, gravimetric soil water content, and N_2O flux in Experiment 1. As earthworm density increased, N_2O flux also increased; however, flux was only significantly higher in the high density treatment at 25% soil water content and at both the medium and high earthworm densities at 35% moisture. This could be attributed to optimal gravimetric soil water content for earthworm activity between 25 and 30%, which closely corresponds to the 31% moisture value reported by the response surface analysis in which emissions were also the highest.

Experiment 2 showed no relationship between earthworm density and N_2O emission, which was expected because the earthworm densities used in Experiment 2 are within the range of the control and low density treatments used in Experiment 1 in which there was no significant difference in N_2O emissions. As a result, the results found here would only have implications in a TBI system where earthworm populations were triple to what is found at GARS. However, earthworms prefer environments with higher organic matter content and soil water content, both of which are present in a TBI system. This could result in higher emissions indirectly related to earthworm population as TBI systems have more favorable environments to earthworms. However, N_2O emissions as a result of the presence of earthworms could be dependent on the proximity of earthworm to the tree row, as well as the species of trees present in the TBI system.

The results of this study are important to consider when deciding on the implementation of agricultural practices to reduce N_2O emissions and also the invasion of earthworms into areas void of earthworms. The benefits that are normally seen from earthworms in agricultural systems may be masked by their influence on facilitating the production of N_2O and in turn, climate change.

KEYWORDS

- **Denitrification**
- **Earthworm density**
- **Earthworm mortality**
- **Mesocosm**
- **Nitrification**
- **Vermicomposting**

ACKNOWLEDGMENT

The study was financed by the Natural Sciences and Engineering Research Council of Canada *via* a Strategic Grant to Dr. Joann Whalen, Dr. Andrew Gordon and others. The authors would like to thank Dr. Joann Whalen from McGill University and Dr. Rick Bourbonniere from Environment Canada for all of their technical assistance.

REFERENCES

1. Rizhiya, E., Bertora, C., Van Vliet, P. C. J., Kuikman, P. J., Faber, J. H., and Van Groenigen, J. W. Earthworm activity as a determinant for N_2O emission from crop residue. *Soil Biology and Biochemistry*, **39**(8), 2058–2069 (2007).
2. Edwards, C. A. and Bohlen, P. J. *Biology and Ecology of Earthworms*, 3rd ed. Chapman and Hall, London, UK (1996).
3. Amador, J. A., Gorres, J. H., and Savin, M. C. Effects of Lumbricus terrestris L. on nitrogen dynamics beyond the burrow. *Applied Soil Ecology*, **33**(1) 61–66 (2006.)
4. Baker, G. H., Brown, G., Butt, K., Curry, J. P., and Scullion, J. Introduced earthworms in agricultural and reclaimed land: their ecology and influences on soil properties, plant production and other soil biota. *Biological Invasions*, **8**(6), 1301–1316 (2006).
5. Cortez, J., Billes, G., and Bouch´e, M. B. Effect of climate, soil, type and earthworm activity on nitrogen transfer from a nitrogen-15-labelled decomposing material under field conditions. *Biology and Fertility of Soils*, **30**(4), 318–327 (2000).
6. Drake, H. L. and Horn, M. A. Earthworms as a transient heaven for terrestrial denitrifying microbes: a review. *Engineering in Life Sciences*, **6**(3), 261–265 (2006).
7. International Panel on Climate change, Fourth Assessment Report AR4, IPCC (2007).
8. Karsten, G. R. and Drake, H. L. Denitrifying bacteria in the earthworm gastrointestinal tract and *in vivo* emission of nitrous oxide (N_2O) by earthworms. *Applied and Environmental Microbiology*, **63**(5), 1878–1882 (1997).
9. Horn, M. A., Schramm, A., and Drake, H. L., The earthwormgut: an ideal habitat for ingested N_2O-producing microorganisms. *Applied and Environmental Microbiology*, **69**(3), 1662–1669 (2003).
10. Horn, M. A., Mertel, R., Gehre, M., Kastner, M., and Drake, H. L. *In vivo* emission of dinitrogen by earthworms *via* denitrifying bacteria in the gut. *Applied and Environmental Microbiology*, **72**(2), 1013–1018 (2006).
11. Price, G. W. and Gordon, A. M. Spatial and temporal distribution of earthworms in a temperate intercropping system in southern Ontario, Canada. *Agroforestry Systems*, **44**(23), 141–149 (1999).

12. Reynolds, P. E., Simpson, J. A., Thevathasan, N. V., and Gordon, A. M. Effects of tree competition on corn and soybean photosynthesis, growth, and yield in a temperate tree-based agroforestry intercropping system in southern Ontario, Canada. *Ecological Engineering*, **29**(4), 362–371 (2007).
13. Price, G. *Spatial and temporal distribution of earthworms (Lumbricidae) in a temperate intercropping system in southern Ontario*, M. S. dissertation, Department of Environmental Biology, University of Guelph, Guelph, Canada (1999).
14. Peichl, M., Thevathasan, N. V., Gordon, A. M., Huss, J., and Abohassan, R. A. Carbon sequestration potentials in temperate tree-based intercropping systems, southern Ontario, Canada. *Agroforestry Systems*, **66**(3), 243–257 (2006).
15. Dymond, P., Scheu, S., and Parkinson, D. Density and distribution of Dendrobaena octaedra (Lumbricidae) in aspen and pine forests in the Canadian Rocky Mountains (Alberta). *Soil Biology and Biochemistry*, **29**(34), 265–273 (1997).
16. Carter, M. R. and Gregorich, E. G. *Soil Sampling and Methods of Analysis*. Canadian Society of Soil Science, Lewis, New York, USA (2008).
17. Kellman, L. and Kavanaugh, K. Nitrous oxide dynamics in managed northern forest soil profiles: is production offset by consumption?. *Biogeochemistry*, **90**(2), 115–128 (2008).
18. Bowley, S. *A Hitchhiker's Guide to Statistics in Plant Biology*, Any Old Subject Books, Guelph, Canada (1999).
19. Bertora, C., Van Vliet, P. C. J., Hummelink, E. W. J., and Van Groenigen J. W. Do earthworms increase N_2O emissions in ploughed grassland?. *Soil Biology and Biochemistry*, **39**(2), 632–640 (2007).
20. El-Duweini, A. K. and Ghabbour, S. I. Nephridial systems and water balance of three Oligochaeta genera. *Oikos*, **19**, 61–70 (1968).
21. Smith, K. A., Thomson, P. E., Clayton, H., McTaggart, I. P., and Conen, F. Effects of temperature, water content, and nitrogen fertilisation on emissions of nitrous oxide by soils, *Atmospheric Environment*, **32**(19), 3301–3309 (1998).
22. Perreault, J. M. and Whalen, J. K. Earthworm burrowing in laboratory microcosms as influenced by soil temperature and moisture. *Pedobiologia*, **50**(5), 397–403 (2006).
23. Schmidt, O. and Curry, J. P. Population dynamics of earth-worms (Lumbricidae) and their role in nitrogen turnover in wheat and wheat-clover cropping systems. *Pedobiologia*, **45**(2), 174–187 (2001).
24. Elliott, P. W., Knight, D., and Anderson, J. M. Variables controlling denitrification from earthworm casts and soil in permanent pastures. *Biology and Fertility of Soils*, **11**(1), 24–29 (1991).
25. Frederickson, J. and Howell, G. Large-scale vermicomposting: emission of nitrous oxide and effects of temperature on earthworm populations. *Pedobiologia*, **47**(56), 724–730, (2003).
26. Wust, P. K., Horn, M. A., and Drake, H. L. *In situ* hydrogen and nitrous oxide as indicators of concomitant fermentation and denitrification in the alimentary canal of the earthwormLumbricus terrestris. *Applied and Environmental Microbiology*, **75**(7), 1852–1859 (2009).
27. Wust, P. K., Horn, M. A., Henderson, G., Janssen, P. H., Rehm, B. H. A., and Drake, H. L. Gut-associated denitrification and *in vivo* emission of nitrous oxide by the earthworm familiesmegascolecidae and lumbricidae in New Zealand. *Applied and Environmental Microbiology*, **75**(11), 3430–3436 (2009).
28. Groffman, P. M., Bohlen, P. J., Fisk, M. C., and Fahey, T. J. Exotic earthworm invasion and microbial biomass in temperate forest soils. *Ecosystems*, **7**(1), 45–54 (2004).
29. Fisk, M. C., Fahey, T. J., Groffman, P. M., and Bohlen, P. J. Earthworm invasion, fine-root distributions, and soil respiration in north temperate forests. *Ecosystems*, **7**(1), 55–62 (2004).
30. Bohlen, P. J., Groffman, P. M., Fahey, T. J., et al. Ecosystem consequences of exotic earthworm invasion of north temperate forests. *Ecosystems*, **7**(1), 1–12 (2004).

10 Bacterial Communities' Response to Nitrogen, Lime, and Plants

Deirdre C. Rooney, Nabla M. Kennedy, Deirdre B. Gleeson, and Nicholas J. W. Clipson

CONTENTS

10.1 INTRODUCTION

The impact of anthropogenic activities on soil biodiversity is central to our understanding of the links between soil functional diversity, species diversity, and overall ecosystem functioning. Agricultural improvement of natural upland pastures is widespread in NW Europe, with increased fertilization, liming, and grazing producing a shift in the floristic composition of acidic upland grasslands [1, 2]. Such intensification practices result in a gradual shift from a plant species-rich *Agrostis capillaris* pasture to a species-poor grassland dominated by *Lolium perenne* [3], with concurrent changes in soil physicochemical properties [1, 4, 5], most notably soil nutrient status. Nitrogen pools in particular have been shown to be held in different ratios between unimproved and improved grasslands, with ammonium dominating unimproved pastures, while nitrate is prevalent in improved pastures [5].

Soil bacterial and fungal communities are also understood to undergo changes in response to agricultural management [4-7], so it is likely that specific functional groups, such as ammonia-oxidizing bacteria (AOB), may be similarly affected. The initial step in nitrification—the conversion of ammonium to nitrite—is microbially mediated by ammonia oxidizers *via* the enzyme ammonia monooxygenase (AMO). In recent years, exploitation of the amoA gene as a molecular marker and the application of community fingerprinting techniques have revealed considerable AOB diversity [8].

While studies have suggested that AOB community structure may be affected by land management regimes [9-11], the impacts of individual components of improvement (i.e., fertiliser addition, liming, and floristic composition) and their interactions on AOB communities in grasslands are not well defined. This study investigates the responses of AOB to nitrogen and lime additions in upland grassland microcosms. Preceding work has indicated that the general rhizosphere bacterial community in acidic grasslands is impacted more by chemical treatments, particularly liming, than by plant species [6]. We hypothesise that the AOB community will be similarly unaffected by plant species. Due to the crucial role of AOB in nitrogen cycling, we expect that nitrogen manipulations will have a greater influence than liming.

10.2 MATERIALS AND METHODS

Soil was collected from an area of unimproved Nardo-Galion grassland at Longhill, County Wicklow, Ireland, and used for microcosms as described [6]. Pots were filled with 80 g (dry mass) of sieved bulk soil, then planted with 20–25 seeds (Emorsgate Seeds, Kings Lynn, UK) of the chosen grassland species (*Agrostis capillaris* or *Lolium perenne*), and water content was adjusted to 35% (w/w). A set of unplanted pots was also included, and microcosms were harvested 75 days after visible germination of seeds. Both planted and unplanted pots were treated on Day 25 as follows: (1) no treatment (NT); (2) addition of lime equivalent to 5 tons ha^{-1} (L); (3) addition of NH_4NO_3 equivalent to 150 kg N ha^{-1} (N); and (4) addition of both lime and NH_4NO_3 as LN. Microcosms were destructively sampled on Day 75. Plant root and shoot material were removed, dried at 70°C for 7 days, and weighed. Due to the small size of the microcosms used and the high root density in pots, all soil was assumed to have been in contact with plant roots and was considered rhizosphere. Soil was sieved to < 4 mm and stored at 4°C for less than 7 days for pH, microbial activity, and biomass analysis, and at for molecular analyses. Soil pH, nitrogen, phosphorus, and potassium were measured as described [6]. Total microbial activity was measured as triphenylformazan dehydrogenase activity [12] and was determined based on a modification of the method of Thalmann [13]. Total soil DNA was extracted as described by Brodie et al. [5]. Briefly, soil (0.5 g) was added to tubes containing glass and zirconia beads, to which CTAB (hexadecyltrimethylammonium bromide) extraction buffer was added. After incubation at 70°C for 10 min, phenol : chloroform : isoamyl alcohol (25 : 24 : 1) was added and tubes were then shaken in a Hybaid Ribolyser at 5.5 m/s for 30 s. Following bead beating, tubes were centrifuged and the aqueous layer was removed and extracted twice with chloroform : isoamyl alcohol (24 : 1). A further purification procedure was performed involving incubation with lysozyme solution (100 mg/mL) for 30 min at 37°C. Tubes were again centrifuged and the aqueous layer removed and further purified using a High Pure PCR Product Clean Up Kit (Roche Diagnostics GmbH, Penzberg, Germany) according to manufacturer's instructions. The DNA was eluted in a final volume of 50 μL. The terminal restriction fragment (TRF) length polymorphism (TRFLP) was carried out using the primer set amo-1F and amo-2R [14, 15], with the forward primer labelled with fluorescent dye

D4. The PCR reactions were performed in 50 μL volumes using PCR Master-Mix (Promega), 15 pmol of each primer, and ~10 ng extracted DNA, and subsequently purified using a PCR product purification kit (Roche). Approximately 50 ng of PCR product was digested using restriction endonuclease TaqI (NEB) according to manufacturer's instructions [15]. One microlitre of desalted digests was mixed with 38.75 μL of sample loading solution and 0.25 μL of Beckman Coulter size standard 600. The TRF lengths were determined by electrophoresis using a Beckman Coulter CEQ 8000 automated sequencer, and analyzed using the Beckman Coulter fragment analysis package v 8.0. Profiles were generated for each sample based on relative heights (abundance) of peaks. Peak heights for each TRF were first converted into proportions of the total peak height of all TRFs for each replicate. TRFs that differed by less than 0.5 bp were considered identical and only fragments occurring in two or more replicates were used [16]. Each fragment was then considered a unique operational taxonomic unit (OTU).

The experimental design consisted of two factors: management type (4 levels, fixed) and plant type (3 levels, fixed) with four replicates. Data sets for soil chemistry, root and shoot biomass, and microbial activity and number of TRFs were analyzed by one-way factorial analysis of variance (ANOVA) using Genstat v 6 (significance level: $p < .05$), after being tested for normality. Multivariate statistical analyses were performed on amoA TRFLP profiles with Primer 6 package (Primer-E Ltd, UK), using standardised and transformed ($\text{Log}(X + 1)$) TRFLP profiles and the Bray-Curtis similarity measure. Principal coordinate analysis (PCO) was used as an unconstrained ordination method to visualise multivariate patterns within each plant treatment. Analysis of similarity (ANOSIM) was used to investigate the effects of treatment and plant species on AOB community structure. The ANOSIM was performed on Bray-Curtis similarity matrices. The comparisons of mean distances within treatments were used to calculate the ANOSIM R-statistic (R), with $R + 1$ indicating the populations were dissimilar, and $R = 0$ indicating that populations were random. Permutational multivariate analysis of variance (PERMANOVA) was also carried out to investigate the effects of treatment and plant species on AOB community profiles, as this procedure can determine if interactions between treatment and plant species were significant. The SIMPER procedure (similarity percentage analysis) was used to identify those TRFs (OTUs) that characterized each treatment group (NT, L, N, LN) identified by cluster analysis [17].

10.3 DISCUSSION

The main finding of this study was that nitrogen addition to upland grassland soil significantly influenced AOB community structure, and that the response was also dependent on which plant species was present. It has been shown before that nitrogen manipulations can lead to shifts in AOB community structure [19, 20], but the simultaneous effects of different plant species on AOB community structure have not been characterized. The results of this study present evidence for floristic and chemical interactions in grasslands selecting for distinct AOB populations. The ability of key functional communities in soils to respond to environmental influences suggests that overall functioning, in this case nitrification may persist over a broad range of ecological conditions.

In terms of plant growth responses to nitrogen and lime treatments in this study, *L. perenne* accounted for the majority of variance seen in root and shoot biomass measurements [6, 7], probably due to fundamental differences in growth morphology and root architectures between the two key plant species. *L. perenne* responds positively to N inputs in soil and it has been suggested that the root system of *L. perenne* may be competitively advantageous over other plant species when high concentrations of soil N are present [21]. In addition, high inputs of fertilizer N is required to maintain the floristic characteristics of improved pastures, in which *L. perenne* dominates. The differences in the fundamental morphologies and physiologies of these two key plant species in response to nitrogen and lime manipulation, may result in concurrent shifts in rhizosphere microbial ecology, including AOB populations.

Liming of grasslands is routinely carried out during agricultural improvement; therefore, AOB community structure may also be influenced by lime-induced pH changes that are concurrent with improvement. Slightly higher soil pH induced by liming may favor certain microbial processes, hence, encouraging greater soil microbial activity. Figure 1 indicates that the presence of *A. capillaris* prevented the LN treatment from substantially altering AOB community structure in these microcosms, when compared to the unplanted control. Additionally, AOB populations under N and LN treatments on the *A. capillaris* ordination plot revealed similar AOB community composition. Interestingly, in both unplanted and *L. perenne* pots, AOB populations in N and LN treatments are markedly different, suggesting differential selection for AOB populations depending on interacting environmental factors. Several studies have indicated that selection for acid/alkaline tolerant AOB species may occur in natural environments in response to shifts in soil pH status [22-24]. As A. capillaris is known to have an acidifying effect on its rhizosphere [25], perhaps this phenomenon prevents lime from influencing soil pH in *A. capillaris* pots as compared to unplanted and *L. perenne* pots (Table 1). The results presented herein propose that liming and nitrogen applications to acidic grasslands may alter the structure of resident AOB populations, selecting for those that are better suited to the prevailing environmental conditions, such as variations in N availability (fertilized/unfertilized) or changes in pH, either as a result of liming or *via* addition of NH_4NO_3, which can acidify soil. In unimproved acidic grasslands, where there is higher floristic and chemical diversity [11], soil heterogeneity could lead to greater grassland AOB diversity. The AOB diversity has been shown to be lower in improved grasslands [11, 26], while greater AOB diversity in unimproved soils is likely to reflect natural physical and chemical heterogeneity, emphasising the importance of soil physicochemical parameters in the determination of microbial community structure and functioning in general.

For unplanted and planted (both *A.capillaris and L. perenne*) soils, there was a high similarity in TRFLP profiles under NT and L treatments (Figure 1), suggesting that lime application alone did not markedly change the AOB community structure in these microcosms. Conversely, the N is proposed as a dominant factor governing AOB community structure in this study, which supports preceding studies [19, 20, 27-29]. Mendum and Hirsch [19] further suggested that NH_4NO_3 application is specifically required for certain *Nitrosospira* clusters to become the dominant AOB population, perhaps indicating a substrate-driven selection for AOB populations, while Cavagnaro et al. [29] suggested that AOB in an agricultural soil was determined by NH_4^+ availability, which selected for ammonia oxidizers that rapidly oxidize available NH_4^+. Additionally, in our study, NH_4NO_3 could alternatively have induced AOB populations to respond to a change in pH, selecting for AOB populations that have adapted to more acidic conditions, as brought about by NH_4NO_3 fertilization. This is further supported by Figure 2, which showed marked differences in AOB population structure under different treatments. Many recent studies on AOB communities have also found that AOB in terrestrial environments are more likely to be dominated by *Nitrosospira* species [30, 31], rather than *Nitrosomonas*. Extensive phylogenetic analyses of β-proteobacteria AOB from terrestrial environments have revealed the presence of a number of clusters within the AOB community, encompassing both Nitrosomonas and Nitrosospira genera [22, 32, 33]. It is possible that these phylogenetic clusters are a result of physiological differences in AOB populations that may allow certain populations to become dominant under different environmental conditions [20], but problems with culturing AOB make it difficult to prove this for certain.

While only two plant species were investigated in this study, natural acidic grasslands maintain a higher diversity of plant species, presenting a highly complex system in which it is difficult to generalize the effects of plant species. Additionally, the Crenarchaea have recently been revealed as important contributors to ammonia oxidization in terrestrial systems [33]. Although crenarchaeal ammonia oxidation was not investigated as part of this study, it is likely that AOB is only partially responsible for regulating ammonia oxidation in grassland soils, so caution must be expressed when evaluating the importance of findings such as those presented in this study. Nevertheless, AOB community dynamics are an integral component of overall ammonia oxidation in soil.

10.4 RESULTS

The soil used for these microcosms had a starting pH of 4.0–4.5. Results of microcosm soil physicochemical analysis are presented in Table 1 (modified from [6]). Soil pH varied significantly ($P < .005$) with chemical treatment, most notably with a significant increase in soil pH after liming (to approximately a value commonly found in semi-improved grassland [5]), and a decrease in soil pH under N treatment (although more acidic than the original soil, this value was still within the normal range for an unimproved acidic grassland [5]). Addition of nitrogen typically reduces soil pH, as ammonium is converted to nitrate (nitrification) [18]. Amendment with lime plus nitrogen raised pH to nearly the same level as lime alone, indicating that the neutralizing effect of lime overrode the acidifying effect of nitrogen. Plant species also altered pH, with lowest pH recorded in *A. capillaris* pots. Plant biomass varied with chemical treatment, with N resulting in higher root biomass across plant treatments, while liming appeared to have the biggest effect on shoot biomass. Plant biomass was significantly different ($P< .001$) depending on plant species, with *L. perenne* producing significantly greater ($P< .001$), root, and shoot biomass compared to *A. capillaris*. Microbial activity decreased significantly under N treatment, whereas lime application (both lime alone, plus lime with N) appeared to have a positive effect on microbial activity. Within plant treatments, a significant decrease ($P< .001$) in microbial activity was noted in *A. capillaris* pots. Soil nitrogen (%), phosphorus and potassium values are also presented in Table 1.

TABLE 1 Soil, plant, and microbial characteristics for grassland microcosms. Nitrogen/Lime treatments, $n = 12$; plant treatments, $n = 16$. Different letters indicate values that are significantly different. SED, standard error of the difference. Significance levels: **$p < .01$; ***$p < .001$. Parts of this table have been formerly presented in the work by Kennedy et al. [6].

	pH	Root Biomass (mg per plant)	Shoot Biomass (mg per plant)	Microbial Activity (mg TPF g^{-1} soil)	Nitrogen (%)	Phosphorus (mg P kg^{-1})	Potassium (mg K kg^{-1})
Nitrogen/Lime treatments							
No treatment (plant only)	4.23b	1.90a	1.94a	49.2b	0.62a	8.54a	140.7b
Lime	4.92d	1.64a	3.68c	158.0d	0.55a	11.88c	156.3d
NH_4NO_3	3.78a	3.81d	2.83b	8.7a	0.62a	9.84b	143.7c
Lime + NH_4NO_3	4.55c	2.72c	4.66d	109.9c	0.60a	12.67d	128.2a
SED	0.09	0.41	0.39	7.05	0.04	0.15	2.20
Plant species							
Unplanted soil	4.50b	N/A	N/A	101.9b	0.61b	7.88a	131.2b
A. capillaris	4.13a	0.79a	0.84a	38.3a	0.48a	11.96b	174.2c
L. perenne	4.48b	4.25b	5.73b	104.2b	0.70c	12.36c	121.4a
P-value	**	***	***	***	***	***	***
SED	0.09	0.49	0.48	11.47	0.04	0.13	2.85

The ANOSIM R-statistics are presented in Table 2. The AOB community structure was significantly affected by both chemical treatment (global R = 0.684) and plant species (global R = 0.584), with significance level at $P < .005$. The larger R values for chemical treatment indicated that treatment had a greater effect on AOB community structure than plant species type. PERMANOVA pair wise comparisons for AOB community structure showed significant effects ($P < .001$) of all management treatments (L, N, LN), plant species, and their interactions on AOB community structure. In order to elucidate the main drivers of AOB community structure, PCO ordinations (Figure 1) were derived and showed marked separation of samples depending on chemical treatment. For both unplanted and planted soils, there was a general grouping together of NT and L treatments, suggesting a high similarity within TRFLP profiles under NT and L treatments. Figure 1(a) indicated that for unplanted soils, the addition of N selected for markedly different AOB communities in comparison to NT and L treatments, while the LN interaction also resulted in a divergent AOB community, indicating a significant combined effect of N and L within these microcosms. Figure 1(b) showed that in *A. capillaris* planted soils, although the NT and L treatments were grouped together and considerably separated from N and LN, there was no definite difference between N and LN, as was seen in the unplanted pots. Figure 1(c) *(L. perenne)* followed a similar pattern to unplanted, with N and LN separating out from each other on the PCO ordination.

TABLE 2 The ANOSIM R-statistics for pairwise comparisons of chemical treatments (lime, nitrogen, lime + nitrogen) and plant species (unplanted, *A. capillaris*, *L. perenne*) on ammonia-oxidizing bacterial communities. NT, no treatment. All comparisons were significant ($P < 0.005$).

Treatment/Plant groups	***R*-statistic**
NT, Lime	0.382
NT, Nitrogen	0.819
NT, Lime + Nitrogen	0.708
Lime, Nitrogen	1
Lime, Lime + Nitrogen	0.852
Nitrogen, Lime + Nitrogen	0.535
Unplanted, *A. capillaris*	0.568
Unplanted, *L. perenne*	0.477
A. capillaris, *L. perenne*	0.755

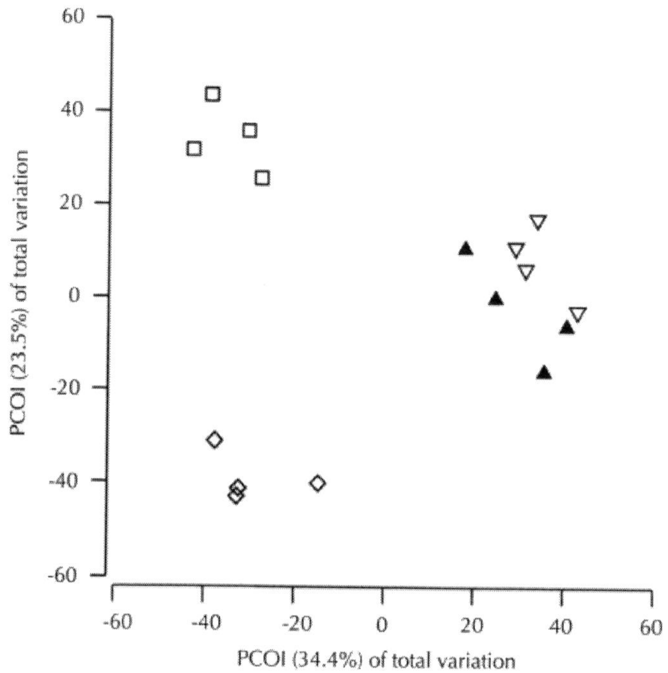

(a)

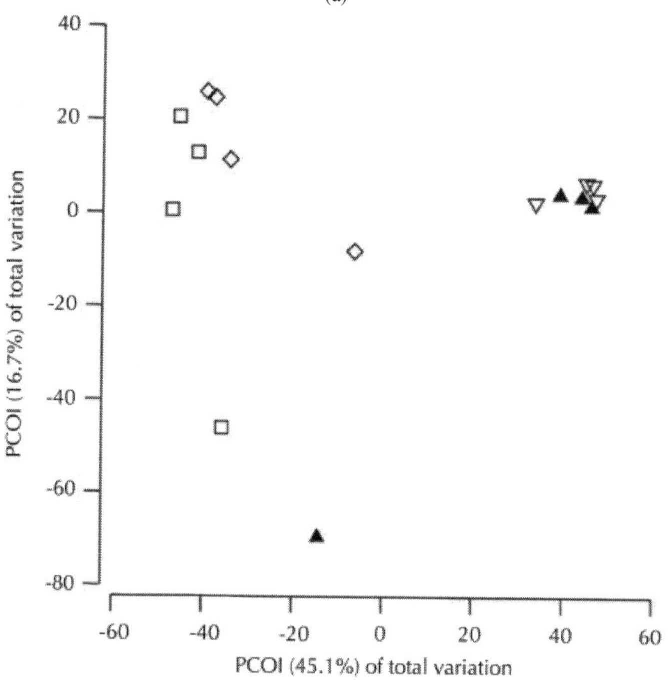

(b)

FIGURE 1 *(Continued)*

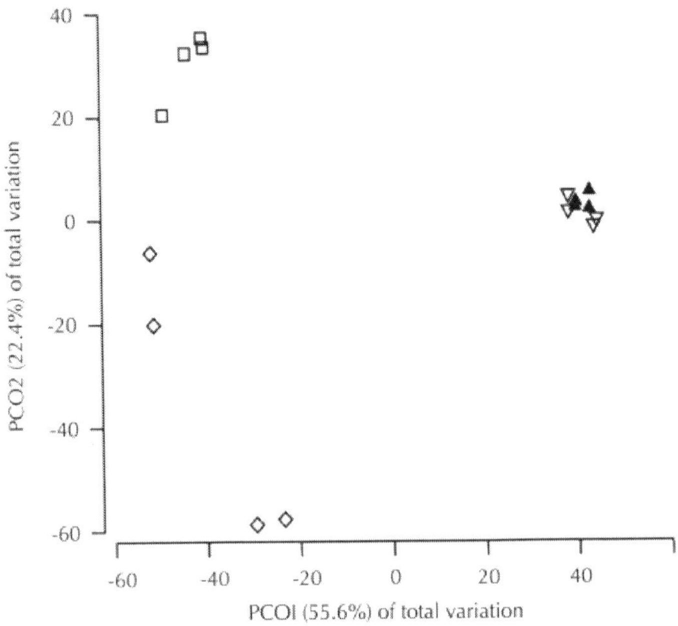

(c)

FIGURE 1 The PCO plots of ammonia-oxidising bacterial community profiles examining management effects within each plant treatment. (a) unplanted soil, (b) *A. capillaris*, (c) *L. perenne*. Key: closed triangle, no treatment; inverted triangle, lime treatment; square, nitrogen treatment; and diamond, lime plus nitrogen treatment.

Figure 2 indicated broad treatment-driven shifts in AOB community structure, as determined from relative abundances of dominant AOB OTUs using the SIMPER procedure. The most abundant OTUs in N-treated microcosms were markedly different from NT and L pots, suggesting apparent selection for different AOB species depending on management factor. Interestingly, the four OTUs present in N treatment (Figure 2) were also present in LN treatment, suggesting that N was very influential in determining AOB community structure. On the other hand, NT and L treatments were more similar in terms of AOB species composition. These data imply that N was dominant in driving AOB community structure. Table 3 gives the number of TRFs present under each treatment. Under NT, there was a slight but significant decrease in number of TRFs detected. There were no significant differences in number of TRFs between plant treatments in the L treated pots, while L, N, and LN treatments revealed some significant differences in TRF number. The most noticeable difference in number of TRFs was in the N treatment and LN treatments, where a significant reduction in number of TRFs in *A. capillaris* and *L. perenne* pots, respectively, was found.

TABLE 3 Number of ammonia oxidizer TRFs (amoA) detected under each treatment and plant species. Different letters represent significant differences within a treatment, n = 4.

	No treatment	Lime	Nitrogen	Lime + Nitrogen
Unplanted	30a	42a	65a	57a
A. capillaris	30a	49a	3c	49a
L. perenne	20b	49a	10b	29b

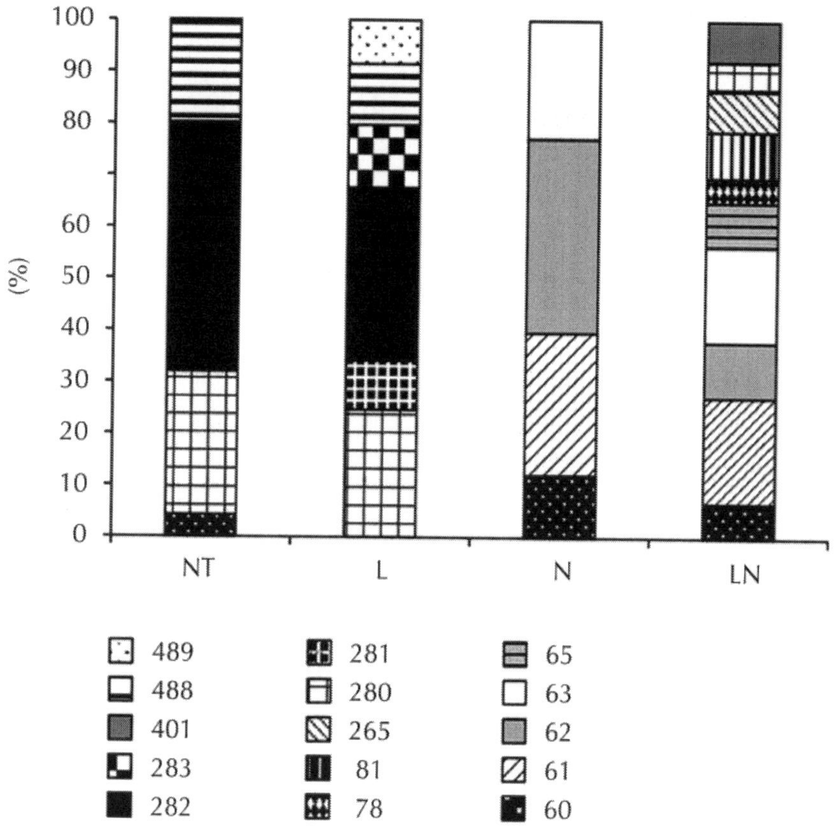

FIGURE 2 Relative percentage abundance of individual ammonia-oxidizing bacterial OTUs occurring in the Top 10 most abundant OTUs within treatment. Top 10 most abundant OTUs were determined based on ranked relative abundances of OTUs within each treatment. No treatment (NT), lime (L), nitrogen (N) and lime + nitrogen (LN) treatments. Abundances are illustrated according to the key at bottom, with the corresponding TRF length (bp) listed alongside.

10.5 CONCLUSION

This study has shown that nitrogen status and plant species type, plus their interactions, play major roles in the determination of AOB community structure in upland acidic pastures. While much valuable data have been presented on the responses of soil AOB communities to changes in soil N status (e.g., [11, 28, 34-36]), often studies do not consider the combined effects of plant species type an important component of agricultural improvement. The data presented herein suggest that individual plant species may affect AOB communities differently in response to fertilizer treatment, adding to the complexity of grassland microbial ecology. Future work should aim to link ammonia-oxidizer community dynamics with soil properties and functional processes, including those regulated by plants, to gain a more comprehensive insight into the factors regulating ammonia oxidizer structure and nitrification dynamics in agricultural systems.

KEYWORDS

- **Ammonia-oxidizing bacteria**
- **Ammonia monooxygenase**
- **Isoamyl alcohol**
- **Nitrosomonas**
- **Terminal restriction fragment**

ACKNOWLEDGMENT

This study was funded by the Environmental Protection Agency under the ERTDI scholarship scheme (2003). The authors thank Mary Murphy and Maria Benson for technical assistance.

REFERENCES

1. Blackstock, T. H., Rimes, C. A., Stevens, D. P., Jefferson R. G., Robertson H. J., Mackintosh, J., and Hopkins, J. J. The extent of semi-natural grassland communities in lowland England and Wales: a review of conservation surveys 1978–1996. *Grass and Forage Science*, **54**(1), 1–18 (1999).
2. Green, B. H. Agricultural intensification and the loss of habitat, species and amenity in British grasslands, a review of historical change and assessment of future prospects. *Grass and Forage Science*, **45**, 365–372 (1990).
3. Rodwell, J. S. *British Plant Communities, Grasslands and Montane Communities*, Cambridge University Press, Cambridge, UK (1992).
4. Bardgett, R. D., Mawdsley, J. L., Edwards, S., Hobbs P. J., Rodwell J. S., and Davies, W. J. Plant species and nitrogen effects on soil biological properties of temperate upland grasslands. *Functional Ecology*, **13**(5), 650–660 (1999).
5. Brodie, E., Edwards, S., and Clipson, N. Bacterial community dynamics across a floristic gradient in a temperate upland grassland ecosystem. *Microbial Ecology*, **44**(3), 260–270 (2002).
6. Kennedy, N., Brodie, E., Connolly, J., and Clipson, N. Impact of lime, nitrogen and plant species on bacterial community structure in grassland microcosms. *Environmental Microbiology*, **6**(10), 1070–1080 (2004).

7. Kennedy, N., Connolly, J., and Clipson, N. Impact of lime, nitrogen and plant species on fungal community structure in grassland microcosms. *Environmental Microbiology*, **7**(6), 780–788 (2005).
8. Prosser, J. I. and Embley, T. M. Cultivation-based and molecular approaches to characterisation of terrestrial and aquatic nitrifiers, *Antonie van Leeuwenhoek*, **81**(1–4), 165–179 (2002).
9. Phillips, C. J., Paul, E. A., and Prosser, J. I. Quantitative analysis of ammonia oxidizing bacteria using competitive PCR, *FEMS Microbiology Ecology*, **32**(2), 167–175 (2000).
10. Ibekwe A. M., Kennedy, A. C., Frohne, P. S., Papiernik, S. K., Yang, C. H., and Crowley, D. E. Microbial diversity along a transect of agronomic zones. *FEMS Microbiology Ecology*, **39**(3), 183–191 (2002).
11. Webster, G., Embley, T. M., and Prosser, J. I. Grassland management regimens reduce small-scale heterogeneity and species diversity of β-proteobacterial ammonia oxidizer populations. *Applied and Environmental Microbiology*, **68**(1), 20–30 (2002).
12. Alef, K. and Nannipieri, P. *Methods in Applied Soil Microbiology and Biochemistry*. Academic Press, London, UK (1995).
13. Thalmann, A. Zur methodik ber Bestimmung der dehydrogenaseaktivität im Boden mittels triphenyltetrazoliumchlorid (TTC), *Landwirtsch Forsch*, **21**, 249–258 (1968).
14. Rotthauwe, J. H., Witzel, K. P., and Liesack, W. The ammonia monooxygenase structural gene amoa as a functional marker: molecular fine-scale analysis of natural ammonia-oxidizing populations. *Applied and Environmental Microbiology*, **63(**12), 4704–4712 (1997).
15. Horz, H. P., Rotthauwe, J. H., Lukow, T., and Liesack, W. Identification of major subgroups of ammonia-oxidizing bacteria in environmental samples by T-RFLP analysis of amoA PCR products. *Journal of Microbiological Methods*, **39(**3), 197–204 (2000).
16. Dunbar, J., Ticknor, L. O., and Kuske, C. R. Phylogenetic specificity and reproducibility and new method for analysis of terminal restriction fragment profiles of 16S rRNA genes from bacterial communities. *Applied and Environmental Microbiology*, **67**(1), 190–197 (2001).
17. Clarke, K. R. Non-parametric multivariate analyses of changes in community structure. *Australian Journal of Ecology*, **18**(1), 117–143 (1993).
18. Fenton, G. and Helyar, K. R. Soil acidification, in Soils: Their Properties and Management, P. E. Charman and B. W. Murphy (Eds.). Oxford University Press, Victoria, Australia, pp. 221–245 (2000).
19. Mendum, T. A. and Hirsch, P. R. Changes in the population structure of β-group autotrophic ammonia oxidizing bacteria in arable soils in response to agricultural practice. *Soil Biology and Biochemistry*, **34**(10), 1479–1485 (2002).
20. Webster, G., Embley, T. M., Freitag, T. E., Smith, Z. and Prosser, J. I. Links between ammonia oxidizer species composition, functional diversity and nitrification kinetics in grassland soils. *Environmental Microbiology*, **7**(5), 676–684 (2005).
21. Personeni, E., Lüscher, A., and Loiseau, P., Rhizosphere activity, grass species and N availability effects on the soil C and N cycles. *Soil Biology and Biochemistry*, **37**(5), 819–827 (2005).
22. Stephen, J. R., McCaig, A. E., Smith, Z., Prosser, J. I., and. Embley, T. M. Molecular diversity of soil and marine 16S rRNA gene sequences related toβ-subgroup ammonia-oxidizing bacteria. *Applied and Environmental Microbiology*, **62**(11), 4147–4154 (1996).
23. Kowalchuk, G. A., Stephen, J. R., de Boer, W., Prosser, J. I., Embley, T. M., and Woldendorp, J. W. Analysis of ammonia-oxidizing bacteria of the β subdivision of the class proteobacteria in coastal sand dunes by denaturing gradient gel electrophoresis and sequencing of PCR-amplified 16S ribosomal DNA fragments. *Applied and Environmental Microbiology*, **63**(4), 1489–1497 (1997).
24. Regan, J. M., Harrington, G. W., and Noguera, D. R. Ammonia- and nitrite-oxidizing bacterial communities in a pilot-scale chloraminated drinking water distribution system. *Applied and Environmental Microbiology*, 68, **1**, 73–81 (2002).
25. Grime, J. P. Benefits of plant diversity to ecosystems: immediate, filter, and founder effects. *Journal of Ecology*, **86**(6), 902–910 (1998).

26. Bruns, M. A., Stephen, J. R., Kowalchuk, G. A., Prosser, J. I., and Paul, E. A. Comparative diversity of ammonia oxidizer 16S rRNA gene sequences in native, tilled, and successional soils. *Applied and Environmental Microbiology*, **65**(7), 2994–3000 (1999).
27. Kowalchuk, G. A., Stienstra, A. W., Heilig, G. H. J., Stephen, J. R., and Woldendorp, J. W. Molecular analysis of ammonia-oxidizing bacteria in soil of successional grasslands of the Drentsche A (Netherlands). *FEMS Microbiology Ecology*, **31**(3), 207–215 (2000).
28. Avrahami, S., Liesack, W., and Conrad, R. Effects of temperature and fertilizer on activity and community structure of soil ammonia oxidizers. *Environmental Microbiology*, **5**(8), 691–705 (2003).
29. Cavagnaro, T. R., Jackson, L. E., Hristova, K., and Scow, K. M. Short-term population dynamics of ammonia oxidizing bacteria in an agricultural soil. *Applied Soil Ecology*, **40**(1), 13–18 (2008).
30. Bäckman, J. S. K., Hermansson, A., Tebbe, C. C., and Lindgren, P. E. Liming induces growth of a diverse flora of ammonia-oxidizing bacteria in acid spruce forest soil as determined by SSCP and DGGE. *Soil Biology and Biochemistry*, **35**(10), 1337–1347 (2003).
31. Mahmood, S. and Prosser, J. I. The influence of synthetic sheep urine on ammonia oxidizing bacterial communities in grassland soil. *FEMS Microbiology Ecology*, **56**(3), 444–454 (2006).
32. Stephen, J. R., Kowalchuk, G. A., Bruns, M. A. V., McCaig, A. E., Phillips, C. J., Embley, T. M., and Prosser, J. I. Analysis of β-subgroup proteobacterial ammonia oxidizer populations in soil by denaturing gradient gel electrophoresis analysis and hierarchical phylogenetic probing. *Applied and Environmental Microbiology*, **64**(8), 2958–2965 (1998).
33. Leininger, S., Urich, T., and Urich, T. Archaea predominate among ammonia-oxidizing prokaryotes in soils. *Nature*, **442**(7104), 806–809 (2006).
34. Avrahami, S. and Bohannan, B. J. M. N_2Oemission rates in a California meadow soil are influenced by fertilizer level, soil moisture and the community structure of ammonia-oxidizing bacteria. *Global Change Biology*, **15**(3), 643–655 (2009).
35. Cantera, J. J. L., Jordan, F. L., and Stein, L. Y. Effects of irrigation sources on ammonia-oxidizing bacterial communities in a managed turf-covered aridisol. *Biology and Fertility of Soils*, **43**(2), 247–255 (2006).
36. Nyberg, K., Schnürer, A., Sundh, I., Jarvis, A., and Hallin, S. Ammonia-oxidizing communities in agricultural soil incubated with organic waste residues. *Biology and Fertility of Soils*, **42**(4), 315–323 (2006).

11 Earthworms, Soil Properties, and Plant Growth

Kam-Rigne Laossi, Thibaud Decaëns, Pascal Jouquet, and Sébastien Barot

CONTENTS

11.1 INTRODUCTION

Earthworms are usually assumed to enhance plant growth through different mechanisms which are now clearly identified. It is however difficult to determine their relative importance, and to predict a priori the strength and direction of the effects of a given earthworm species on a given plant. Soil properties are likely to be very influential in determining plant responses to earthworm activities. They are likely to change the relative strength of the various mechanisms involved in plant earthworm interactions. We review the different rationales used to explain changes in earthworm effect due to soil type in this chapter. Then, we systematically discuss the effect of main soil characteristics (soil texture, OM, and nutrient contents) on the different mechanisms allowing earthworm to influence plant growth. Finally, we identify the main shortcomings in knowledge and point out the new experimental and meta-analytical approaches that need to be developed. An example of such a meta-analysis is given and means to go further are suggested. The result highlights a strong positive effect size in sandy soil and a weakly negative effect in clayey soil.

Earthworms are among the most important detritivores in terrestrial ecosystems in terms of biomass and activity [1]. They are known to affect plant growth through five main mechanisms [2, 3]: (1) the enhancement of soil organic matter mineralization, (2) the production of plant growth regulators *via* the stimulation of microbial activity, (3) the control of pests and parasites, (4) the stimulation of symbionts, and (5) the modifications of soil porosity and aggregation, which induce changes in water and oxygen availability to plant roots. Although these mechanisms are well identified, it is difficult to determine their relative influence [4] and to predict the impact of a given earthworm species on a given plant species. In a recent review, Brown et al. [2] proposed that the response of plants to earthworms should depend on soil properties such as texture, mineral nutrient levels, and organic matter content. However, most studies tackling earthworm effects on plant growth used soils containing more sand than clay (Brown et al. [2] and see Table 1). Comparatively, few studies [5-7] have tested in the same experiment earthworm effects on plant growth using different soils. Doube et al. [5] showed that the endogeic Aporrectodea trapezoides may increase wheat growth in sandy soils but may have no significant effect with a clayey substrate. They also found that the growth and grain yield of barley were both increased by A. trapezoides and Aporrectodea rosea in the sandy soil but reduced in the clayey one. On the contrary, Laossi et al. [7] showed that Lumbricus terrestris increased the shoot and total biomasses of *Trifolium dubium* in a clayey and nutrient-rich soil but not in a sandy and nutrient-poor one. The hypothesis that earthworm effects on plant growth should vary with soil type is based on two main reasons. (1) Soil properties may inhibit or stimulate some of the mechanisms through which earthworms tend to increase plant growth. (2) If earthworms are able to alleviate limiting factor for plant growth, their impacts are expected to be weak in soils where the factor is not limiting. According to this rationale, the main mechanism through which earthworms affect plants should depend on soil type and in some soils earthworms might have no detectable or negative effect on plant growth.

11.2 HOW SOIL PROPERTIES SHOULD MODULATE EARTHWORM EFFECTS ON PLANT GROWTH?

We go through the different mechanisms listed and try to determine how soil properties should modulate their effect on plant growth.

(1) Earthworm activities usually have a positive impact on the mineralization of soil organic matter [8]. This effect is assumed to be a consequence of plant litter fragmentation and incorporation into the soil, as well as of the selective stimulation of microbial activity [9, 10]. Hence earthworms may enhance the release of nutrients that become available to plants and thus increase plant growth when they allow higher nutrient uptake than nutrient leaching [11, 12]. Anecic and endogeic earthworms have different feeding habits and affect differently soil organic matter composition and distribution [13]. Anecic earthworms feed on plant litter at the soil surface and tend to live in semipermanent vertical burrows while endogeic earthworms are active within the soil profile where they feed on soil organic matter [14]. This can lead to different effect on plant growth [15-17] which could also vary with soil properties such as organic matter and nutrient contents [18]. However, this rationale only holds if nutrients are limiting plant growth that is in soils where nutrients are poorly

TABLE 1 References included in the survey of C and N contents in soil and the soil texture used in earthworm effects on plant growth.

References	Soil classification	Soil texture	N total	Clay	Sand	C Total
Aira and Plearce 2009	John Innes potting compost no. 2	Loamy compost	?	?	20%	?
Blouin et al. 2005	Ultisol	Sandy soil	0.05%	6%	78%	0.91%
Blouin et al. 2006	Ultisol	Sandy soil	0.05%	6%	78%	0.91%
Blouin et al. 2007	Ultisol	Sandy soil	0.05%	6%	78%	0.91%
Bonkowski et al. 2001	?	Loam soil + sand	0.1%	?	>50%	1.52%
Clapperton et al. 2001	Chernozem	Loamy soil	?	?	?	?
Derouard et al. 1997	?	Sandy soil	0.11%	5%	87%	0.91%
Devliegher and Verstraete 1997	Ardoyne	Sandy soil	?	?	?	0.9%
Daube et al. 1997	Xerosol, Palexeralf, wiesenboden	Sandy soil, loamy soil and clayey soil	?	?	?	?
Eisenhauer et al. 2009	Eutric Fluvisol	Loam soil	0.3%	?	?	4.6%
Eisenhauer et al. 2008	Gleyic cambisol	Silty soil	?	22%	9%	1.1%
Eisenhauer and Scheu 2008a	Eutric Fluvisol	Loam soil	0.3%	?	?	4.6%
Eisenhauer and Scheu 2008b	Eutric Fluvisol	Loam soil	0.3%	?	?	4.6%
Eriksen-Hamel and Whalen 2007	Typic endoquent	Sandy loam soil	?	12%	58%	2.45%
Eriksen-Hamel and Whalen 2008	Typic endoquent	Loam soil	?	?	?	5.6%–13.3%
Fraser et al. 2003	Udic dystrochrept	Silt loam soil	0.2%	?	?	2.6%

TABLE 1 *(Continued)*

References	Soil classification	Soil texture	N total	Clay	Sand	C Total
Gilot 1997	Ferralsol	Sandy soil	0.44%–0.59%	6%–10%	>75%	>1%
Gilot-Villenave et al. 1996	Ferralsol	Sandy soil	0.4%–1.2%	2.4%–4.5%	>80%	0.28%–1.18%
Hale et al. 2008	Eutroboralf	Silty clay loam soil	?	?	?	?
Hale et al. 2006	Eutroboralf	?	?	?	?	?
Hopp and Slatter 1948	?	Clayey soil	?	70%	16%	?
Kreuzer et al. 2004	?	?	<0.1%	?	?	1.8%
Taossi et al. 2009a	Cambisol	Sandy soil	0.12%	6.9%	74%	1.47%
Taossi et al. 2009b	Cambisol and leptosol	Sandy soil and clayey soil	0.12%; 0.46%	6.9%; 34.4%	74%; 27%	1.47%; 5.67%
Milcu et al. 2006	Eutric Fluvisol	Loam soil	0.3%	?	?	4.6%
Milleret et al. 2009	Anthrosol	Sandy soil + compost	?	26.7%	45.3%	?
Newington et al. 2004	?	Sandy loam + aquatic compost +leaf mulch	<0.01%	?	?	?
Ortiz-Ceballos et al. 2007a	Fluvisol	Silty clay loam soil	0.25%	26.8%	41.5%	?
Ortiz-Ceballos et al. 2007b	Fluvisol	Silty clay loam soil	0.25%	26.8%	41.5%	?
Partsch et al. 2006	Eutric Fluvisol	Loam soil	0.3%	?	15%	4.6%
Pashanasi et al. 1996	Paleudult	Sandy loam soil	?	23%	55%	?

TABLE 1 *(Continued)*

References	Soil classification	Soil texture	N total	Clay	Sand	C Total
Patron et al. 1999	Inceptisol	Sandy soil	0.08%	11%	82%	0.85%
Poveda et al. 2005a	?	?	?	?	?	?
Poveda et al. 2005b	?	?	?	?	?	?
Schmidt and Curry 1999	Podzol	Loam to clay loam soil	0.18%	19%	47%	1.88%
Stephens and Davoren 1995	Calcic Natrixeralf	Calcic soil	?	?	?	1.5%
Stephens and Davoren 1997	Calcic Natrixeralf	Calcic soil	?	?	?	?
Thompson et al. 1993	?	Loam soil	?	?	?	?
Welke and Parkinson 2003	Dystric brunisols, gray-brown luvisols	Sandy loam soil	0.1%–1.07%	?	?	21.63%–43.73%
Wurst et al. 2008	?	Loamy sandy mineral soil	0.13%	?	?	2.1%
Wurst et al. 2005	Cambisol	Loam soil	0.087%	?	?	1.58%
Wurst et al. 2003	?	Loam soil	0.087%	?	?	1.58%
Zaller and Arnone 1999	Rendzina	Calcareous soil	?	?	?	?

available. In contrast, in nutrient-rich soils, plants are less limited by the availability of mineral nutrients and earthworm-mediated mineralization should have less or no influence on plant growth [2]. Water is between the factors that limit plant growth and earthworms have been found to increase drought stress in plants [19]. This effect should be stronger in sandy soil, which retains less water than in a clayey one.

(2) Earthworms affect plant growth through modifications of soil structure. They tend to increase soil porosity and the stability of organomineral aggregates by creating burrows and organomineral casts at different places within the soil profile [20, 21]. This effect is assumed to enhance plant growth in most situations [2] although opposite effects have also been reported [22]. It is difficult to predict how soil texture will modulate these effects. In clayey soils, earthworm might lead to very stable structures which could in turn strongly influence plant growth. This influence could be positive if the casts produced by earthworms do not lead to soil compaction [22], or negative with a physical protection of organic matter that impedes the release of mineral nutrients. In sandy soils, structures created by earthworms are more fragile [23] but more mineral nutrients can be released since the soil organic matter is less protected.

(3) Earthworm effects on plant growth *via* the release of plant growth regulators may be modulated by soil properties through several mechanisms, but here again the outcome is difficult to be predicted.
First, plant growth regulators are thought to be released by bacteria [24] and may be differently available depending on the levels of microbial activity in the soil. Sandy soils and soil with low organic matter contents usually have lower microbial biomasses and low potential for plant growth regulator production [25]. Thus, in such soils, earthworm effects *via* production of plant growth regulator could lead to weak effects on plant growth. Second, soil texture and soil organic matter could also affect the short-term availability of the produced phytohormones. For instance, clays and organic matter are known to adsorb organic molecules [26] and could reduce plant growth regulator availability to plants and weaken earthworm effect on plant growth.

(4) Earthworms are known to alleviate the negative effect of some parasites on plant growth by reducing strongly their density [27], ingesting and killing some pathogens in their intestine, or producing unfavorable conditions in cast material or tunnel lining [28]. This kind of mechanism may be influential for plant growth, especially in soil properties (such as moisture and temperature) that allow the development of abundant parasite populations. We can thus expect more parasites and greater negative effect of earthworms on them in a clayey soil.

(5) Similarly, earthworms can increase plant growth through the stimulation of symbionts or the increase in the contact between plants and symbionts [29]. Besides, if symbionts such as mycorrhizae provide nutrients to plants, symbiont-mediated earthworm effect (as their effect through mineralization) on plants should be more marked in poor soils than in rich soils where mineral resources are already available. Taken together, these elements show that earthworm

effects on plants vary with soil type but that it is difficult to predict the direction and the intensity of these variations. To make relevant predictions, we need to develop studies comparing in the same experiment earthworm effects on plants under different soil conditions. It is also necessary to set up meta-analyses using data of earthworms—plants studies. We provide an example of what could be done through computing the effect size of earthworms on plant growth using meta-analysis with the data of the studies listed in Table 1.

11.2.1 How Can We Go Further?

To determine how earthworms effects on plant growth change with soil properties a first approach would be to compare earthworm-induced effects in different soils but in the same experimental conditions (same plant and earthworm species, same watering protocol, same greenhouse, etc.). Such experiments have been so far very scarce (but see [5-7]). To help predicting earthworm effects on plant growth in different soil types one could also use the "all-minus-one" tests proposed by Brown et al. [2]. In such experiments, only one factor such as mineral nutrition [4] or a root parasite [27] is limiting plant growth so that the capacity of earthworms to alleviate this limiting factor can be tested. This allows testing main mechanisms through which earthworms affect plant growth in particular conditions. Such experiments could be repeated in soils differing by only one property to determine how this property modulates the strength of each of these mechanisms. For example, the experimental studies conducted in greenhouse conditions [27, 30] and using sandy soil have showed that earthworms enhanced the tolerance of plants to nematodes. This kind of study should be carried out using sandy and clayey soil in the same experiment to test whether soils properties change the strength and direction of this earthworm effect.

While comparing earthworm effect in different soils differing by only one parameter is easy, this is not likely to allow disentangling all factors because soil properties are often correlated. Clay soils are generally rich in organic matter. A solution would be to directly manipulate soil properties. Hence, it would be possible to add, for example, clay, sand, organic matter, or mineral nutrients to a soil. We would then study the effect of a gradient in clay, sand, organic matter, or mineral nutrient on earthworm-induced effect. In the same vein, earthworm effect on the nutrient input-output balance of ecosystems should determine the long-term effect of earthworm on plant primary production [31]. Thus, comparing earthworm effects in different soils could also allow measuring their effects on nutrient leaching in these different soils and to identify the type of soil in which nutrients made available are leached and in which other it remains in the superficial soil layers. This is important to determine the long-term effect of earthworm soil type interaction on plant growth. Another possibility would be to conduct meta-analyses to take advantage of the numerous studies.

11.3 EXAMPLE OF META-ANALYSIS

We used the results of 25 experiments (Table 1) to perform a meta-analysis and calculate the effect size of earthworms influence on plant growth in different soil types with contrasting texture properties (sandy, clayey, or loamy soil). The effect size was computed as (M1-M2)/σ, with M1: mean plant biomass in the presence of earthworms,

M2: mean plant biomass without earthworm, and σ: standard deviation without earthworm [32]. An analysis of variance (ANOVA) was then used to test for the effect of soil texture on the effect size that is on the magnitude of earthworm impact on plant biomass. This shows that soil texture influences significantly earthworm impact on plant growth ($r^2 = 0.11$ and $P = 0.02$). The LS mean comparisons show that the effect size was in sandy and loamy soils, respectively, 60 and 17.5 fold higher than in clayey soil. The result highlights a strong positive effect size in sandy soil and a weakly negative effect in clayey soil (Figure 1).

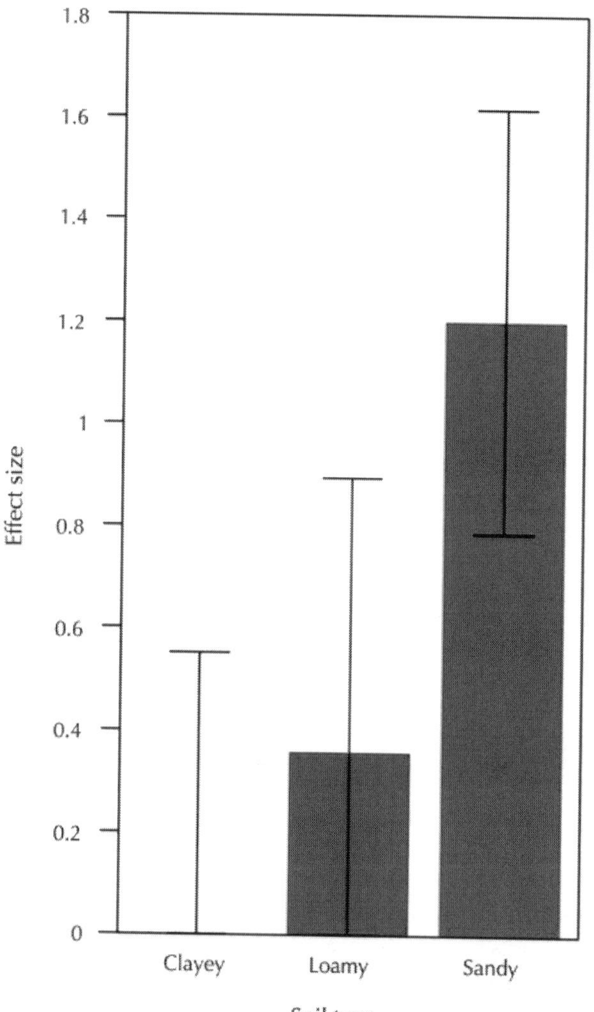

FIGURE 1 Effect size of earthworm effects on plant growth based on results of 25 experiments presented in Table 1. Mean effect size calculated as [M1-M2]/σ with M1: mean plant biomass in the presence of earthworms, M2: mean plant biomass without earthworm, and σ: standard deviation without earthworm. P = 0.02.

This result supports the assertion of Brown et al. [2] that positive effects of earthworms on plant growth are more pronounced in sandy soils (generally nutrient-poor soils) that in clayey soils (generally nutrient-rich soil). However, as showed in Table 1 most studies used sandy soils while only few studies have used clayey ones. We thus need to release this bias by developing more studies for clayey soil. Nevertheless, the meta-analysis is the first formal test of the influence of soil properties on earthworm effect on plant growth.

11.4 CONCLUSION

Although the majority of authors provided detailed data on soil characteristics, this basic information was not available in all studies in earthworm impacts on plant production (Table 1). Further studies should pay attention to providing a standardized description of soil characteristics which would thus be available for meta-analyses on earthworms—plants studies. For example, data on soil texture (sand and clay percentage), total C, total N content, NH_4^+, and NO_3^- should be systematically published. Because such information, is not always given (see Table 1), we have only compared the effect of wide texture classes on earthworm effect. Finally, we have shown that these texture classes only explain 11% of the variations in effect sizes. This is probably due to a variety of other factors that we have not taken into account: soil properties mentioned but also earthworm species (or its functional group) and plant species (or its functional group), and so forth [2, 3]. Gathering more studies on earthworm effects on plant growth and documenting for each of these studies all these factors would allow disentangling, through a unique meta-analysis statistical model, the respective effect of all these factors on earthworm-induced effect on plant growth, as well as interactions between these factors. This kind of general and systematic approach is required to derive general results on soil ecology and to develop the theoretical background needed to base soil ecology on solid bases [33]. Taken together, while a given earthworm species could have positive effects in a soil, it could have negative effects in another soil. To restore soil fertility or to enhance the sustainability of crop production [14], the right earthworm species has indeed to be chosen according to soil properties and crop type. Developing applications based on the use of earthworms would thus also require implementing the general meta-analysis and the subsequent development of a general and comprehensive framework on earthworm-induced effect on plant growth that is so far missing.

KEYWORDS

- **Meta-analyses**
- **Nutrient-rich soil**
- **Plant growth regulator**
- **Soil organic matter**
- **Soil texture**

REFERENCES

1. C. A. Edwards (Ed.). *Earthworm Ecology*. CRC Press, Boca Raton, Florida, USA (2004).
2. Brown, G. G., Edwards, C. A., and Brussaard, L. How earthworms effect plant growth: burrowing into the mechanisms. In: *Earthworm Ecology*, C. A. Edwards (Ed.), pp. 13–49 (2004).
3. Scheu, S. Effects of earthworms on plant growth: patterns and perspectives. *Pedobiologia*, **47**(5–6), 846–856 (2003).
4. Blouin, M., Barot, S., and Lavelle, P. Earthworms (Millsonia anomala, Megascolecidae) do not increase rice growth through enhanced nitrogen mineralization. *Soil Biology and Biochemistry*, **38**(8), 2063–2068 (2006).
5. Doube, B. M., Williams, P. M. L., and Willmott, P. J. The influence of two species of earthworm (Aporrectodea trapezoides and Aporrectoedea rosea) on the growth of wheat, barley and faba beans in three soil types in the greenhouse. *Soil Biology and Biochemistry*, **29**(3–4), 503–509 (1997).
6. Wurst, S. and Jones, T. H. Indirect effects of earthworms (Aporrectodea caliginosa) on an above-ground tritrophic interaction. *Pedobiologia*, **47**(1), 91–97 (2003).
7. Laossi, K. R., Ginot, A., Noguera, D. C., Blouin, M., and Barot, S. Earthworm effects on plant growth do not necessarily decrease with soil fertility. *Plant and Soil*, **328**(1–2), 109–118 (2010).
8. Bohlen, P. J., Pelletier, D. M., Groffman, P. M., Fahey, T. J., and Fisk, M. C. Influence of earthworm invasion on redistribution and retention of soil carbon and nitrogen in northern temperate forests. *Ecosystems*, **7**(1), 13–27 (2004).
9. Bohlen, P. J., Edwards, C. A., Zhang, Q., Parmelee, R. W., and Allen, M. Indirect effects of earthworms on microbial assimilation of labile carbon. *Applied Soil Ecology*, **20**(3), 255–261 (2002).
10. Martin, A., Mariotti, A., Balesdent, J., and Lavelle, P. Soil organic matter assimilation by a geophagous tropical earthworm based on δ13C measurements. *Ecology*, **73**(1), 118–128 (1992).
11. Domínguez, J., Bohlen, P. J., and Parmelee, R. W. Earthworms increase nitrogen leaching to greater soil depths in row crop agroecosystems. *Ecosystems*, **7**(6), 672–685 (2004).
12. Jouquet, P., Bernard-Reversat, F., Bottinelli, N., et al. Influence of changes in land use and earthworm activities on carbon and nitrogen dynamics in a steepland ecosystem in Northern Vietnam. *Biology and Fertility of Soils*, **44**(1), 69–77 (2007).
13. Bouché, M. B. Stratégies lombriciennes. In: *Soil Organisms as Components of Ecosystems*, U. Lohm and T. Persson (Eds.). Ecological Bulletins 25, Stockholm, Sweden, pp. 122–132 (1977).
14. Lavelle, P., Barois, I., Blanchart, E., et al. Earthworms as a resource in tropical agroesosystems. *Nature and Resources*, 34, 26–40 (1998).
15. Laossi, K. R., Noguera, D. C., and Barot, S. Earthworm-mediated maternal effects on seed germination and seedling growth in three annual plants. *Soil Biology and Biochemistry*, **42**(2), 319–323 (2010).
16. Laossi, K. R. *Effet des vers de terre sur les plantes: du fonctionnement individuel à la structure des communautés végétales*. Ph.D. thesis. Université Pierre et Marie Curie, Paris, France, 2009.
17. Laossi, K. R., Noguera, D. C., Mathieu, J., Blouin, M., and Barot, S. Barot Effects of an endogeic and an anecic earthworm on the competition between four annual plants and their relative fecundity. *Soil Biology and Biochemistry*, **41**(8), 1668–1673 (2009).
18. Jouquet, P., Dauber, J., Lagerlöf, J., Lavelle, P., and Lepage, M. Soil invertebrates as ecosystem engineers: intended and accidental effects on soil and feedback loops. *Applied Soil Ecology*, **32**(2), 153–164 (2006).
19. Blouin, M., Lavelle, P., and Laffray, D. Drought stress in rice (Oryza sativa L.) is enhanced in the presence of the compacting earthworm Millsonia anomala. *Environmental and Experimental Botany*, **60**(3), 352–359 (2007).
20. Lavelle, P., Decaëns, T., Aubert, M., et al. Soil invertebrates and ecosystem services. *European Journal of Soil Biology*, **42**(1), S3–S15(2006).
21. Kretzschmar, A. Effects of earthworms on soil organization. In: *Earthworm Ecology*, C. A. Edwards (Ed.). Taylor & Francis, London, UK, 201–210 (2004).

22. Chauvel, A., Grimaldi, M., Barros, E., et al. Pasture damage by an Amazonian earthworm. *Nature*, **398**(6722), 32–33 (1999).
23. Lavelle, P. and Spain, A. *Soil Ecology*. Kluwer Academic Publishers, Dordrecht, Netherlands (2001).
24. Ping, L. and Boland, W. Signals from the underground: bacterial volatiles promote growth in Arabidopsis. *Trends in Plant Science*, **9**(6), 263–266 (2004).
25. Hendrix, P. F., Peterson, A. C., Beare, M. H., and Coleman, D. C. Long-term effects of earthworms on microbial biomass nitrogen in coarse and fine textured soils. *Applied Soil Ecology*, **9**(1–3), 375–380 (1998).
26. Huang, Q., Liang, W., and Cai, P. Adsorption, desorption and activities of acid phosphatase on various colloidal particles from an Ultisol. *Colloids and Surfaces B*, **45**(3–4), 209–214 (2005).
27. Blouin, M., Zuily-Fodil, Y., Pham-Thi, A. T., et al. Belowground organism activities effect plant aboveground phenotype, including plant tolerance to parasites. *Ecology Letters*, **8**, 202–208 (2005).
28. Stephens, P. M. and Davoren, C. W. Influence of the earthworms Aporrectodea trapezoides and A. Rosea on the disease severity of Rhizoctonia solani on subterranean clover and ryegrass. *Soil Biology and Biochemistry*, **29**(3–4), 511–516 (1997).
29. Doube, B. M., Ryder, M. H., Davoren, C. W., and Stephens, P. M. Enhanced root nodulation of subterranean clover (Trifolium subterraneum) by Rhizobium leguminosarum biovar trifolii in the presence of the earthworm Aporrectodea trapezoides (Lumbricidae). *Biology and Fertility of Soils*, **18**(3), 169–174 (1994).
30. Wurst, S., Allema, B., Duyts, H., and Van der Putten, W. H. Earthworms counterbalance the negative effect of microorganisms on plant diversity and enhance the tolerance of grasses to nematodes. *Oikos*, **117**(5), 711–718 (2008).
31. Barot, S., Ugolini, A., and Brikci, F. B. Nutrient cycling efficiency explains the long-term effect of ecosystem engineers on primary production. *Functional Ecology*, **21**(1), 1–10 (2007).
32. Hedges, L. V. and Olkin, I. Statistical Methods for Meta-Analysis. Academic Press, New York, USA (1985).
33. Barot, S., Blouin, M., Fontaine, S., Jouquet, P., Lata, J. C., and Mathieu, J. A tale of four stories: soil ecology, theory, evolution, and the publication system. *Plos One*, **2** article e12 (2007).

12 Organic Versus Conventional Strawberry Agroecosystem

John P. Reganold, Preston K. Andrews, Jennifer R. Reeve, Lynne Carpenter-Boggs, Christopher W. Schadt, J. Richard Alldredge, Carolyn F. Ross, Neal M. Davies, and Jizhong Zhou

CONTENTS

12.1 INTRODUCTION

Sale of organic (ORG) foods is one of the fastest growing market segments within the global food industry. People often buy organic food because they believe organic farms produce more nutritious and better tasting food from healthier soils. Here we tested if there are significant differences in fruit and soil quality from 13 pairs of commercial organic and conventional strawberry agroecosystems in California. At multiple sampling times for 2 years, we evaluated three varieties of strawberries for mineral elements, shelf life, phytochemical composition, and organoleptic properties. We also analyzed traditional soil properties and soil DNA using microarray technology. We found that the organic farms had strawberries with longer shelf life, greater dry matter, and higher antioxidant activity and concentrations of ascorbic acid and phenolic compounds, but lower concentrations of phosphorus and potassium. In one variety, sensory panels judged organic strawberries to be sweeter and have better flavor, overall acceptance, and appearance than their conventional counterparts. We also found the organically farmed soils to have more total carbon and nitrogen, greater microbial biomass and activity, and higher concentrations of micronutrients. Organically farmed soils also exhibited greater numbers of endemic genes and greater functional gene abundance and diversity for several biogeochemical processes, such as nitrogen fixation and pesticide degradation.

Although global demand for organic products remains robust, consumer demand for these products is concentrated in North America and Europe [1]. For example, in the United States, which ranks 4th in organically farmed land globally [1], organic food sales have increased by almost a factor of six, from \$3.6 billion in 1997 to \$21.1 billion in 2008 (or more than 3% of total U.S. food sales) [2]. More than two-thirds of U.S. consumers buy organic products at least occasionally, and 28% buy organic products weekly [2]. Three of the most important reasons consumers purchase organic foods are health benefits (i.e., less pesticide residues and greater nutrition), taste, and environmentally friendly farming practices, such as those that promote soil health [3]. While there is strong evidence that organic foods have significantly less pesticide residues [4-6], this is not the case for organic foods being more nutritious. Although there is no universally accepted definition of what constitutes a nutritious food, recent scientific opinion has stressed that more nutritious foods are those that are more nutrient dense relative to their energy contents [7]. Although carbohydrates and fats are considered essential nutrients, the current concept of nutrient dense foods, and hence more nutritious foods, places the emphasis on foods that contain more protein, fiber, vitamins, or minerals, as well as specific phytochemicals, such as the polyphenolic antioxidants found in fruits and vegetables [8]. In the past 10 years, ten review studies of the scientific literature comparing the nutrition of organic and conventional foods have been published. Eight of these review studies [9-16] found some evidence of organic food being more nutritious, whereas two reviews [17, 18] concluded that there were no consistent nutritional differences between organic and conventional foods. Comparisons of foods from organic and conventional systems are often complicated by the interactive effects of farming practices, soil quality, plant varieties, and the time of harvest on nutritional quality. Hence, many of the comparative studies cited in some of the earlier reviews were not experimentally well designed to draw valid conclusions [13, 17]; for example, soils or crop varieties were not the same on each organic/conventional field pair. The few studies that have compared organic and conventional foods for their organoleptic (sensory) properties have shown mixed results or used unreliable experimental designs [14, 17]. A widely accepted definition of soil quality is the capacity of a soil to sustain biological productivity, maintain environmental quality, and promote plant and animal health [19]. Soil quality may be inferred from measurable soil properties termed soil quality indicators [20]. Organic farming practices compared to conventional farming practices have been shown to improve soil quality indicators based on traditional measures of biological, chemical, and physical properties [21-23], with few studies showing no advantages [24]. However, traditional measures inadequately assess the roles of microbial community structure and genetic diversity in soil ecosystem processes that directly impact soil quality [25]. Examples of important soil ecosystem processes facilitated by microorganisms include nitrogen fixation, denitrification, pesticide degradation, and other organic xenobiotic degradation. Soil DNA analysis using microarray technology can target those microbial genes involved in specific soil ecosystem processes and measure their abundance and diversity [26, 27], allowing a more complete investigation of soil quality. The majority of organic/conventional studies have focused on either comparing fruit quality or soil quality. The few studies that have compared both facets have limited their analyses to selected properties. Currently, no study has integrated interdisciplinary knowledge and robust methodologies in a systems approach to quantitatively compare a comprehensive range of both fruit and soil quality indices using multiple ORG and conventional farms, multiple varieties, and multiple sampling times. Here, we assembled an interdisciplinary team of scientists representing agroecology, soil science, microbial ecology, genetics, pomology, food chemistry, sensory science, and statistics to address the following question: Are there significant differences in nutritional and organoleptic fruit properties and in soil quality, including soil ecosystem functional genes, between commercial ORG and conventional strawberry agroecosystems?

Although some farm production conditions can be simulated at research stations, farming systems research that measures multiple variables can often only be properly studied under actual farming or agroecosystem conditions [28]. Thus, the study's experimental units are real commercial ORG and conventional strawberry farms, located in California. We chose to study strawberries (*Fragaria x ananassa Duch.)* as the food of choice because of their high economic value as a fruit crop, documented nutritional benefits, popularity in the consumer diet, and suitability for sensory evaluation. California is an appropriate location for the commercial strawberry farms in study because it is the leading producer, accounting for more than 25% of the world's strawberry production [29, 30], with nearly 5% of its total strawberry acreage in ORG production [29]. To determine if differences in food and soil quality exist, we sampled repeatedly harvested strawberry varieties ("Diamante," "San Juan," and "Lanai") and soils at multiple sampling times in 2004 and 2005 from 13 pairs of adjacent ORG and conventional fields from commercial farms. Each ORG/conventional field pair had the same soil type and the same strawberry variety planted at similar times. Because strawberries go through different growth cycles during the 7 month harvest season, we analyzed 42 fruit, 11 leaf, and 6 organoleptic properties multiple times during the 2 years of study. Strawberries in each field pair were analyzed at the same time and stage of harvest maturity, and under identical storage conditions and transportation methods so that the strawberries were as close to retail consumption as possible. In addition to measuring 31 traditional soil chemical and biological properties, we analyzed soil DNA using microarray technology to target those microbial genes involved in 11 specific ecosystem processes.

12.2 METHODS

12.2.1 Study Area

The 13 pairs of side-by-side commercial ORG and conventional strawberry farm fields were selected in the Watsonville area, the dominant strawberry growing region of California, USA. In turn, California is the leading producer in the U.S., accounting for 87% of the nation's strawberry production [29]. The Watsonville area annually grows strawberries on about 5,000 ha, accounting for about 40% of the strawberry acreage in the state (29).

The selection of 13 field pairs (5 in 2004 and 8 in 2005) from commercial strawberry farms was made on the basis of grower interviews and on-farm field examinations to ensure that all soil-forming factors, except management, were the same for each field pair [21]. Each field pair consisted of two side-by-side fields, one ORG and one conventional. Fields chosen in each pair had the same microclimate, soil profile, soil type, soil classification, and strawberry variety.

Strawberry field pairs in 2004 were different from those in 2005 because both ORG and conventional farmers grew strawberries in alternate years using a similar 2 year rotation. More specifically, all farmers in the study grew strawberries on constructed, 30 cm high mounded rows covered with plastic mulch for only 1 year, preceded by a different crop, such as broccoli, lettuce, or a cover crop, grown on flat ground (without mounds) the preceding year. Growing strawberries as annuals using this "raised-bed plasticulture" system is typical of ORG and conventional strawberry growers in California [51]. Growers in the study transplanted strawberry crowns in November, with strawberry plants starting to produce fruit in mid-March and continuing to produce fruit to mid- or end-October.

The ORG fields had been certified organic (United States Department of Agriculture (USDA)) for at least 5 years, providing sufficient time for the ORG farming practices to influence soil properties. The ORG fields relied only on organically certified fertilizers and pesticides and no soil fumigation. Both ORG and conventional farms applied compost, with the ORG strawberry systems using 20.2–24.6 Mg compost per hectare and the conventional systems using 11.2–13.4 Mg compost per hectare. These high rates of compost additions along with ORG fertilizer amendments permitted strawberries to be grown in the 2 year rotation. The conventional farms also had been managed conventionally for at least 5 years and included the use of inorganic and organic fertilizers, synthetic pesticides, and soil fumigation (methyl bromide with or without chloropicrin). Conventional strawberry growers in California typically rely on methyl bromide (with or without chloropicrin) as an extremely effective broad-spectrum, pre-plant biocide to kill soil-borne diseases (including fungi and bacteria), nematodes, soil-dwelling insects, weeds, weed seeds, and underground plant parts [51].

12.2.2 Strawberry Sampling and Analyzes

Strawberry (*Fragaria x ananassa Duch.*) varieties grown on the study farms included "Diamante" and "San Juan" in 2004 and "Diamante," "San Juan," and "Lanai" in 2005. Strawberry fruit were collected from each of 13 pairs of ORG and conventional strawberry farm fields in June and September 2004 and April, June, and September 2005. Commercial pickers harvested and packed ripe fruit in plastic "clamshells," just as they would be sold in retail markets. All strawberries were picked at random from within 8 to 15 rows and were always a minimum of 20 m from the boundary within each field pair to avoid edge effects. Within hours, packed fruit were transported to a refrigerated storage facility until shipped on commercial refrigerated trucks from Watsonville, CA to distribution centers in Seattle or Spokane, WA. These samples were transported under cool conditions to Washington State University (WSU) in Pullman, WA, where they were immediately placed in refrigerated storage.

A subsample of the collected fruit, as well as leaf samples, taken in April 2005 and June 2004 and 2005, were sent to Soiltest Farm Consultants Inc. in Moses Lake, WA, where fruit and leaf samples were analyzed for N, P, K, Ca, Mg, B, and Zn, plus S, Mn, Cu, and Fe for leaf samples only, according to standard methods [52]. All other strawberry analyzes, including shelf life, fresh weight, dry matter, firmness, color, total antioxidant activity, ascorbic acid, total phenolics and anthocyanins, and specific polyphenols, were conducted at Washington State University, Pullman, WA. Since study focused on the nutritional differences of fresh strawberries as consumed, we expressed nutritional composition data on the basis of how the product would be eaten, or fresh weight, which is the standard for the USDA National Nutrient Database for Standard Reference [36], FAO's International Network of Food Data Systems (INFOODS) [53], and the United Kingdom's Food Standards Agency [54].

Strawberry fruit from each field and sampling time were subsampled for fresh analysis within 2 days of receipt at WSU in Pullman, WA, with another subsample stored at −80°C for later biochemical analysis. On each sample time, 20 fruit from each field were weighed fresh, ten of which were dried in an oven at 80°C and reweighed to determine dry weight, while the other 10 fruit were left at room temperature (~20°C) for 2 days and reweighed to estimate weight loss. Fruit firmness was measured as maximum penetration force (N) on opposite sides of another 25 fruit from each field with an automated penetrometer (Model GS-20 Fruit Texture Analyzer, Güss Manufacturing Ltd., Strand, South Africa) fitted with a 5 mm diameter convex cylinder set to a trigger threshold of 1.11 N and 6 mm depth. On each of these fruit, two external (on opposing shoulders) and two internal (adjacent to central cavity) color measurements (Model CR-300 Chroma Meter, Minolta Camera Co., Ltd., Ramsey, NJ) were taken using the L*a*b* color space expressed as lightness (L*), chroma (C*, $[(a^*)^2+(b^*)^2]1/2$) and hue angle (hab, $\tan^{-1}$ [b*/a*]) [55]. Soluble solids content (%) in the homogenate from these berries was measured in triplicate with a digital refractometer (Model PR-101, Atago Co., Ltd., Tokyo, Japan), as were pH and titratable acidity (citric acid equivalents), using an automated titrator (Schott Titroline easy, Schott-Geräte GambH, Mainz, Germany) with 0.1 N KOH (pH 12.8). The ratio of soluble solids to titratable acidity was also calculated.

In order to estimate the susceptibility of the strawberries to fungal rots, a subsample of 72 fresh fruit from each field and sampling time were placed in individual cells (6.7 ×5.9 ×5.7 cm deep) of plastic greenhouse inserts (Model IKN3601, ITML Traditional Series Inserts, Hummert International, Earth City, MO). Two inserts with 36 berries in each were placed in trays with dampened paper to maintain a saturated atmosphere and in sealed, black plastic bags. For both sample months in 2004, fruit were incubated at 15.5°C for 9–10 days, with the number of rotted berries counted each day. All rotted fruit were removed from their cells until all fruit had rotted. The principal fungal rot observed on the berries was gray mold (Botrytis cinerea).

For analysis of antioxidant activity, ascorbic acid, total phenolics, anthocyanins, and total and reducing sugars, chemicals and enzymes were purchased from Sigma-Aldrich Corp. (St. Louis, MO), unless otherwise noted. Spectrophotometric measurements were made using a UV-visible spectrophotometer (Model HP8453, Hewlett-Packard Co., Palo Alto, CA) with UV-visible ChemStation software [Rev. A.08.03(71), Agilent Technologies, Inc., Santa Clara, CA]. All solutions were made up using ultrapure water (NANOpure DIamond Analytical, Barnstead International, Dubuque, IO). Centrifugation was performed in a Eppendorf 5417 R microcentrifuge (Engelsdorf, Germany). There were 3–5 separate replicates of pooled tissue from a minimum of 5 fruit analyzed in each biochemical assay, with duplicate instrument measurements made on each replicate. Outlying data were discarded and the tissue reanalyzed.

Antioxidant activity of hydrophilic and lipophilic fractions [56] in the berries was measured by the end-point 2,2-azino-bis-(3-ethylbenzthiazoline-6-sulfonicacid) (ABTS)/hydrogen peroxide/peroxidase (Horseradish peroxidase, HRP, Type VI-A) method of Cano et al. [57], with modifications. Specifically, 100 mg powdered, frozen berry tissue was extracted in 700 mL of 50 mM MES (pH 6.0) and 700 mL ethyl acetate, vortexed for 30 sec, and centrifuged at 13 K rpm for 10 min at 4°C. The ORG (top) and aqueous (bottom) phases were separated with a pipette for measurement of lipophilic and hydrophilic antioxidant activities (LAA and HAA, respectively). For both fractions, 40 mL of 1 mM H_2O_2, 100 mL of 15 mM ABTS, and 10 mL of 3.3 U mL^{-1} HRP were placed in 1 mL quartz cuvettes and gently shaken for 10 sec, after which 830 mL of 50 mM phosphate buffer (pH 7.4) was added and mixed with a stir paddle. Absorbance was monitored at 734 nm on a UV-visible spectrophotometer until stable (<10 sec), and then 20 mL (for HAA) or 40 mL (for LAA) extract was added, mixed with a stir paddle, and monitored at 734 nm until absorbance reached a minimum. HAA and LAA were calculated from the absorbance difference and expressed on the basis of Trolox equivalents from standard curves of 5 mM Trolox diluted in 50 mM MES buffer (pH 6.0) or 100% (v/v) ethyl acetate, respectively, and measured as described for the samples. HAA and LAA were summed to estimate total antioxidant activity (TAA).

Total ascorbic acid (reduced AsA plus dehydroascorbic acid, DHA) in the berries was measured as originally described by Foyer et al. [58] and modified by Andrews et al. [59]. Specifically, 200 mg powdered, frozen berry tissue was extracted in 1.5 mL ice-cold of 5 M $HClO_4$ by grinding with liquid nitrogen in a mortar and pestle. Samples were transferred into 2 mL brown, microcentrifuge tubes, vortexed for 30 sec, and centrifuged at 13 K rpm for 10 min at 4°C. Into two, 400 mL aliquots of supernatant from each extract, 200 mL of 0.1 M HEPES-KOH buffer (pH 7.0) were added and mixed, followed by 20–30 mL of 5 M K_2CO_3 to reach pH 4–5. Following centrifugation, 200 mL supernatant was reduced by adding 31.8 mL of 1 M DL-dithiothreitol (DTT) in 400 mL of 100 mM phosphate buffer (pH 5.6), gently shaking and incubating on ice for 5 min. Absorbance of 100 mL of reduced extract in 396 mL of 100 mM phosphate buffer (pH 5.6) in a blackened, 0.5 mL quartz cuvette was monitored at 265 nm on a UV-visible spectrophotometer until stable (<10 sec), and then 4 mL of 1 U mL^{-1} ascorbate oxidase (from Cucurbita) was added, mixed with a stir paddle, and monitored at 265 nm until absorbance reached a minimum. Concentration of total ascorbic acid was calculated from the absorbance difference and standard curves of 5.25 mM dehydro-L-(+)-ascorbic acid dimer reduced with 265 mL of 1 M DTT in 400 mL of 100 mM phosphate buffer (pH 5.6) and monitored at 265 nm.

Total phenolic compounds in the berries were measured with the Folin-Ciocalteu (F-C) phenol reagent (2 N) according to revised methods of Singleton et al. [60]. Specifically, to 200 mg powdered, frozen berry tissue, 1 mL of 80% (v/v) methanol was added in microcentrifuge tubes. Samples were vortexed, allowed to extract 1 hr at room temperature and then overnight at −20°C, followed by centrifugation at 14 K rpm for 20 min at 4°C. The supernatants were removed and extraction of the pellet was repeated 2 times as described, with supernatants combined after each extraction and then made up to 4 mL with 80% (v/v) methanol after the final extraction. Total phenolic compounds were assayed by adding 400 mL sample extract into two 15 mL tubes containing 600 mL of 80% (v/v) methanol, 5 mL of 10% (v/v) F-C reagent, and either 4 mL saturated Na_2CO_3 (75 gL^{-1}) or 4 mL water. Tubes were thoroughly mixed and incubated at room temperature for 2 hr. One-milliliter aliquots from the sample tubes containing Na_2CO_3 or water were added to 1.5 mL plastic cuvettes and the absorbance of each was measured at 760 nm in a UV-visible spectrophotometer. Concentration of phenolic compounds was determined by subtracting absorbance of samples containing Na_2CO_3 from those not containing Na_2CO_3, quantified as gallic acid (3,4,5-trihydroxybenzoic acid) equivalents from standard curves.

For anthocyanins, 0.5 g of powdered, frozen berry tissue was extracted in 1 mL of 1% (v/v) HCl-methanol. After storage for 24 hr at −20°C, sample tubes were centrifuged at 14 K rpm for 10 min at 4°C. Extraction with HCl-methanol was repeated 2 times. Following centrifugation on 4th day, supernatants were decanted into 15 mL plastic tubes and made up to 3 mL volumes with HCl-methanol. Anthocyanin concentrations were determined by measuring absorbance of 250 mL extract in 750 mL of 1% (v/v) HCl-methanol in 1.4 mL quartz cuvettes at 515 nm with a UV-visible spectrophotometer [61], expressed as pelargonidyn-3-glucoside equivalents using Emolar = 3.6 × 106 $M^{-1} m^{-1}$. Specific polyphenolic compounds were extracted by grinding 0.1 g frozen, powdered fruit tissue in 1.5 mL pure methanol. Concentrations of aglycones of ellagic acid, quercetin, kaempferol, phloretin, and naringenin enantiomers were determined, as well as the total aglycone plus glycoside polyphenols, following enzymatic hydrolysis with b-glucuronidase from Helix pomatia (Type HP-2) [62, 63]. Extracts (150 mL), with daidzein as internal standard (IS), were injected into a HPLC system (Shimadzu, Kyoto, Japan), consisting of LC-10AT VP pump, SIL-10AF auto injector, SCL-10A system controller. Polyphenols were separated isocratically with a mobile phase of acetonitrile:water:phosphoric acid (v/v/v 42:58:0.01 at 0.6 mL min^{-1} for ellagic acid, quercetin, kaempferol, and phloretin; 30:70:0.04 at 0.4 mL min^{-1} for naringenin enantiomers) at 25°C on chiral stationary phase amylose- or cellulose-coated columns (Chiralcel AD-RH for ellagic acid, quercetin, kaempferol, and phloretin and Chiralcel OD-RH for naringenin enantiomers, with 5 mm particle size and 150 × 4.5 mm ID; Chiral Technologies Inc., Exton, PA, USA), and detected at 370 nm (for ellagic acid, quercetin, kaempferol, and phloretin) and 292 nm (for naringenin enantiomers) on a Shimadzu SPD-M10A VP diode array spectrophotometer. Data collection and peak integration were carried out using Shimadzu EZStart 7.1.1 SP1 software. Individual polyphenols were quantified based on standard curves constructed using peak area ratio (PAR = PApolyphenol/PAIS) against the concentration of the standards. Best laboratory practices during sample analysis followed guidance, based upon the International Conference on Harmonisation for the quantitative analysis of polyphenolic compounds using a validated assay and commercially available standards, with all samples run in duplicate with appropriate quality controls [64, 65].

Reducing and total sugars were measured by the Nelson-Somogyi micro-colorimetric method [66], with modifications. Specifically, 0.1 g frozen berry homogenate was extracted in 1.5 mL pure methanol for 30 min at room temperature, after vortexing for 30 sec. Total sugars were obtained by adding 150 mL of 0.1 M HCl to duplicate tissue samples and allowing hydrolysis of sugars for 10 min prior to methanol extraction. Samples were then centrifuged at 14 K rpm for 10 min. Supernatants (0.2 mL), diluted with 0.8 mL water, were mixed with 1 mL copper-sulfate reagent in glass tubes with stoppers and incubated for 10 min in a boiling water bath. After cooling for 5 min, 1 mL arsenomolybdate reagent was added and mixed. Volumes were adjusted to 10 or 25 mL with deionized water, depending on color density. Concentrations of reducing and total sugars were determined by measuring absorbance at 520 nm with a UV-visible spectrophotometer and quantified by standard curves of glucose made from stock 1% (w/v) glucose solution in saturated benzoic acid.

12.2.3 Sensory Analyzes

We also conducted consumer-sensory analyses of strawberries, including flavor, sweetness, appearance, juiciness, tartness, and overall acceptance. Strawberries were evaluated by consumer-sensory panels at four different sampling dates (20 panelists per field pair in September 2004 and 25 panelists per field pair in April, June, and September 2005) at WSU's Food Science and Human Nutrition Sensory Laboratory. Panelists were recruited using advertising from the WSU community based on their availability. A minimum amount of information on the nature of the study was provided in order to reduce potential bias. All participants signed an Informed Consent Form per project approval by the WSU Institutional Review Board.

Each panelist completed a demographic questionnaire prior to the start of the panel. Fifty-eight percent of panelists were females. The age distribution of the panelists was 31% of 18–25 years old, 41% of 26–35 years old, 10% of 36–45 years old, 13% of 46–55 years old, and <5% over 55 years old. Over 70% of the panelists ate fresh strawberries every 2 weeks to every month, with 19% eating fresh strawberries every week. The majority of panelists (59%) preferred fresh strawberries that tasted more sweet than tart and another 36% preferred them at least equally sweet and tart. Less than 5% of the panelists preferred them more tart than sweet or had no preference.

Each consumer received ORG and conventional berries from two matched field pairs. Consumers were presented with two strawberry halves from two individual strawberries. Each sample was presented in a monadic, randomized serving order with assigned three-digit codes.

Each panelist was also provided with deionized, filtered water and unsalted crackers for cleansing the palate between samples. Consumers evaluated each strawberry sample for overall acceptance as well as perceived intensity of flavor, juiciness, sweetness, and sourness using a discrete 9 point bipolar hedonic/intensity scale, where 1 = dislike extremely/extremely low intensity and 9 = like extremely/extremely high intensity, according to ISO standards for quantitative response scales [67]. These evaluations were completed under red lights to disguise color differences between the samples. Following the taste/flavor evaluations, the lights were changed to white lights and panelists evaluated the strawberries for acceptance of appearance using the same 9 point scale.

12.2.4 Statistical Analyzes of Strawberry Data

Mixed model analyzes of variance were used to test for differences in response variable means, except where noted, due to varieties ("Diamante," "Lanai," and "San Juan"), treatments (ORG and CON), and months (April, June, and September). A split plot model pooled over 2 years was selected with variety as the whole plot factor, treatment as the subplot factor, and month as a repeated measure (SAS Proc Mixed, SAS Institute, 1999). Transformations were used to improve normality and homogeneity of variances where necessary. When data were transformed, LS means were reported in original units. When significant interactions were identified, differences in simple effect means were identified using Fisher's least significant differences. The same mixed model analysis of variance was applied to examine sensory data by using the average panel score for each attribute. The Kaplan-Meier (Product Limit) method was used to model the survival function and estimate mean survival time, that is days to rotting (SAS Proc Lifetest, SAS Institute, 1999). The generalized Savage (Log-Rank) test for equality of survival functions was used to test for differences in time to rotting for organic versus conventional conditions [68].

12.2.5 Soil Sampling and Analyzes

Soils were sampled from 30 cm raised mounds at 0–10 and 20–30 cm depths in June 2004 and June 2005 and at 0–10 cm in April 2005. All samples were a composite of 10–15 subsamples taken at random from within 8 to 15 rows and were always a minimum of 20 m from the boundary within each field pair to avoid edge effects. Samples from the June sampling dates were shipped for chemical analyzes to Soiltest Farm Consultants and for biological analyzes to WSU by overnight mail. Samples from the April 2005 sampling were shipped to Oak Ridge National Lab for microarray analyzes and stored at − 20°C.

At Soiltest Farm Consultants, soil samples were passed through a 2 mm sieve, stored at 4°C, and then analyzed for the following properties according to recommended soil testing methods by Gavlak et al. [52]: Nitrate-nitrogen (N) was measured with the chromotropic acid method; ammonium-N was measured with the salicylate method; Olsen phosphorus was measured; DTPA-Sorpitol extractable sulfur, boron, zinc, manganese, copper, and iron were measured; Soil pH and electrical conductivity were measured in a 1:1 w/v water saturated paste; SMP soil buffer pH was measured; NH_4OAc extractable potassium, calcium, magnesium, and sodium were measured; total bases were calculated by summation of extractable bases; and particle size (percentage sand, silt, and clay) was analyzed by the hydrometer method.

At WSU, we analyzed total C and N by combustion using a Leco CNS 2000 (Leco Corporation, St. Joseph, MI). Readily mineralizable carbon (MinC), basal microbial respiration, and active microbial biomass (MicC) by substrate-induced respiration were measured according to Anderson and Domsch [69]. Ten grams of wet weight soil were brought to 12, 18, and 26% moisture content (0.033 MPa), depending on soil type, and incubated at 24°C for 10 days. Total CO_2 released after 10 days was considered MinC. Vials were recapped for 2 hr and the hourly rate measured for microbial respiration. For MicC, 0.5 mL of 12 g L^{-1} aqueous solution of glucose was added to the same soil samples and rested for 1 hr before being recapped for 2 hr. Carbon dioxide was measured in the headspace using a Shimadzu GC model GC -17A (Shimadzu Scientific Instruments, Columbia, MD), with a thermal conductivity detector and a 168 mm HaySep 100/120 column. From these microbial properties, we calculated the metabolic quotient, qCO_2 (basal respiration/MicC), and the two ratios, MicC/MinC and MicC as a percent of total C. Dehydrogenase enzyme activity was measured using 2.5 g dry weight soil and acid and alkaline phosphatase enzyme activities were measured using 1 g dry weight soil as described by Tabatabai [70]. These enzyme reactions were measured using a Bio-Tek microplate reader model EL311 (Bio-Tek Instruments, Winooski, VT). Both native and potential protease enzyme activities were measured using 1 g dry weight soil according to Ladd and Butler [71] and measured on a Perkin Elmer Lambda 2 UV/VIS spectrometer (PerkinElmer Life And Analytical Sciences, Inc, Waltham, MA) at 700 nm with tyrosine standards. Native protease represents activity without the addition of casein substrate and potential protease represents the activity with the addition of substrate. Arbuscular mycorrhizae were stained with trypan blue and total and colonized roots estimated using the gridline intersection method [72].

12.2.6 Statistical Analyzes of Soils Data

June comparisons of soil under ORG and conventional management were analyzed as a randomized complete block design with split plot. Year served as whole plot and treatment as subplot. The two depth intervals were analyzed separately. All statistics were analyzed using the SAS system for Windows version 9.1 ANOVA and LS means (SAS Institute, 1999). Data were checked for model assumptions and transformed as necessary. When data were transformed, LS means were reported in original units.

12.2.7 Microarray Analyzes

Soil community DNA was extracted using an SDS-based method [73]. A total of 10 g soil from each ORG field and 20 g soil from each conventional field (due to low yields of DNA) were used. DNA was purified using a Wizard PCR cleanup system (Promega,

Madison, WI). The cleaned pellet was washed in 500 μl ethanol (70%) before being re-suspended in 20 μl of 10 mM Tris (pH 8.0). Microarray slides were constructed according to methods described [26]. We used a comprehensive functional gene array, termed GeoChip, containing more than 24,000 oligonucleotide (50 mer) probes and covering 10,000 genes involved in nitrogen, carbon, sulfur, and phosphorus transformations and cycling, metal reduction and resistance, and ORG xenobiotic degradation [41]. Microarray genes were analyzed both individually and as 11 functional groups: nitrification, denitrification, nitrogen fixation, sulfite reduction, pesticide degradation, other ORG xenobiotic degradation, metal reduction and resistance, dehydrogenase, urease, cellulase, and chitinase.

Thirty to 150 ng purified DNA from each soil was randomly amplified using rolling circle PCR with a GenomiPhi DNA amplification kit (GE Healthcare, Piscataway, NJ) [74]. The amplification product was fluorescently labeled with Cy5 dye with an extended 6 hr incubation time and applied directly to the microarray. Mean labeling efficiency per treatment was calculated to ensure no overall bias. Slides and all solutions were kept at 60°C during assembly to minimize cross contamination. Hybridizations were carried out at 50°C with 50% formamide [26]. After hybridization, the slides were immediately placed in wash solution 1 (1 × SSC and 0.1% SDS) to remove the cover slip and washed by gentle shaking in solution 1, 2 times for 5 min each; then washed in solution 2 (0.1 × SSC and 0.1% SDS), 2 times for 10 min each; and finally in solution 3 (0.1 × SSC), 4 times for 1 min each. Arrays were then dried using compressed air. All arrays were run in triplicate. The microarrays were scanned using a ScanArray 5000 analysis system (Perkin-Elmer, Wellesley, MA) [26].

12.2.8 Microarray Data Processing and Analyzes

Microarray slide images were converted to TIFF files and hybridized DNA quantified using ImaGene software 6.0 (Biodiscovery Inc., Los Angeles, CA) [26]. The signal-to-noise ratio (SNR) of each probe on each slide was calculated as follows:

SNR = (signal intensity – local background) (standard deviation of slide background)$^{-1}$. Background refers to the local background intensity, while the standard deviation was calculated across the whole slide. Signal intensity data for any gene was removed unless it appeared with SNR >2 on at least two of three replicate array hybridizations. When this condition was met, individual SNR values< 2 were retained in order to maintain a continuous data set for statistical analysis. The array included multiple probes for some genes; here the strongest signal was retained and weaker ones deleted. After screening for adequate SNR, signal intensity values were then used as the data for sample comparison. Signal intensity values were normalized by averaging across technical replicates and imported into SAS system for Windows version 9.1 ANOVA (SAS Institute, Cary, NC) for analysis.

Data were analyzed using a randomized complete block design with field pair as block. Average signal intensity for each of the 1711 detected genes, sum of signal intensities (SIs) for all 1711 detected genes, and sum of SIs for each of the 11 functional groups from the eight organically farmed soils and the eight matched conventionally farmed soils were analyzed by paired t-tests. Gene diversity was calculated overall and for each functional group using a modified version of Simpson's Reciprocal Index [$D = 1/[\sum n(n-1)/N(N-1)]$, where n = signal intensity of a single gene with an SNR >2 and N = sum of all SIs with an SNR >2 on the entire slide]. Diversity values were then analyzed by a paired t-test. Detected endemic genes were counted based on treatment means. Proportion comparison z tests were used to compare proportion of detected endemic genes in each management system.

12.3 DISCUSSION AND RESULTS

12.3.1 Strawberry Quality

Strawberry leaves were analyzed for plant nutrients and fruit were analyzed for plant nutrients, fruit quality, nutritional value, and organoleptic properties. 1 Leaf P and fruit P and K concentrations were significantly higher in conventionally grown strawberry plants than in organically grown plants (Table 1); leaf Mg and fruit N were also nota-

TABLE 1 Mineral elements (mean ± standard error) in strawberry leaves and fruit from ORG and conventional (CON) farms (n = 13).

Mineral Element	'Diamante'		'Lanai'		'San Juan'		*P* Value
	ORG	CON	ORG	CON	ORG	CON	
			Leaves				
Nitrogen (% DW)	2.51 ± 0.10	2.76* ± 0.10	3.06 ± 0.12	2.91 ± 0.12	2.84 ± 0.11	2.84 ± 0.11	0.020
Calcium (% DW)	0.73 ± 0.36	1.19† ± 0.36	0.81 ± 0.37	0.88 ± 0.37	0.87 ± 0.37	0.77 ± 0.37	0.036
		ORG			CON		
Phosphorus (% DW)		0.37 ± 0.016			0.45 ± 0.016		0.001
Potassium (% DW)		1.56 ± 0.04			1.58 ± 0.04		0.71
Sulfur (% DW)		0.215 ± 0.009			0.214 ± 0.009		0.91
Magnesium (% DW)		0.311 ± 0.047			0.354 ± 0.047		0.066
Boron (ppm DW)		38.9 ± 2.36			38.7 ± 2.36		0.95
Zinc (ppm DW)		59.9 ± 1.31			63.3 ± 1.31		0.73
Manganese (ppm DW)		128 ± 34.5			182 ± 34.5		0.19
Copper (ppm DW)		5.24 ± 1.46			4.81 ± 1.46		0.14
Iron (ppm DW)		207 ± 24.1			214 ± 24.1		0.70
			Fruit				
Nitrogen (% DW)		1.02 ± 0.11			1.08 ± 0.11		0.078
(% FW)		0.105 ± 0.009			0.103 ± 0.009		0.63
Phosphorus (% DW)		0.247 ± 0.012			0.286 ± 0.012		0.001
(% FW)		0.026 ± 0.001			0.027 ± 0.001		0.11
Potassium (% DW)		1.50 ± 0.05			1.65 ± 0.05		0.010
(% FW)		0.154 ± 0.007			0.157 ± 0.007		0.63
Calcium (% DW)		0.120 ± 0.015			0.132 ± 0.015		0.18
(% FW)		0.012 ± 0.002			0.013 ± 0.002		0.89
Magnesium (% DW)		0.130 ± 0.003			0.134 ± 0.003		0.26
(% FW)		0.013 ± 0.0004			0.013 ± 0.0004		0.27
Boron (ppm DW)		14.5 ± 17.5			15.2 ± 17.5		0.57
(ppm FW)		1.53 ± 1.63			1.47 ± 1.63		0.73
Zinc (ppm DW)		9.95 ± 0.71			9.96 ± 0.71		0.99
(ppm FW)		1.03 ± 0.05			0.95 ± 0.05		0.26

Leaves and fruit were sampled in June 2004 and April and June 2005 from 13 pairs of organic (ORG) and conventional (CON) farm fields. Probabilities (*P* values) for treatment x variety interactions for leaf N and Ca, and for treatment main effects for the remaining leaf and fruit mineral elements are given. Means and standard errors of mineral elements in leaves and fruit for individual sampling/harvest times, varieties, and years are listed in Tables S4A, C, and D.

*Means are notably different at $P < 0.10$.

†Means are significantly different at $P < 0.01$.

bly higher (P< 0.10) in conventionally grown strawberry plants. All other strawberry and leaf nutrient concentrations were similar. While there are no recommendations for optimum levels of foliar concentrations of mineral nutrients for strawberries grown in California, all farm fields were fertilized according to local industry standards, as recommended by professional horticulturists. No nutrient deficiency or toxicity symptoms were observed on organically or conventionally grown strawberry plants during the two growing seasons.

When susceptibility to fungal post-harvest rots was evaluated, ORG strawberries had significantly longer survival times (less gray mold incidence) than conventional (CON) strawberries (Figure 1). When strawberries were exposed to a 2 day shelf-life interval, the percent loss in fresh weight was significantly less for the ORG berries than for the CON berries (Table 2). These results indicate that the ORG strawberries would have a longer shelf life than the CON strawberries because of slower rotting and dehydration, perhaps due to augmentation of cuticle and epidermal cell walls. There were no fungicides applied to the ORG strawberry fields for post-harvest control of gray mold (*Botrytis cinerea*), in contrast to multiple fungicide applications to the CON fields. Although sulfur was applied to the ORG fields to control powdery mildew (*Sphaerotheca macularis*), sulfur sprays are ineffective against gray mold [31]. This suggests that the ORG strawberries may have been more resistant or avoided infection by means other than fungicides (e.g., systemic-acquired resistance).

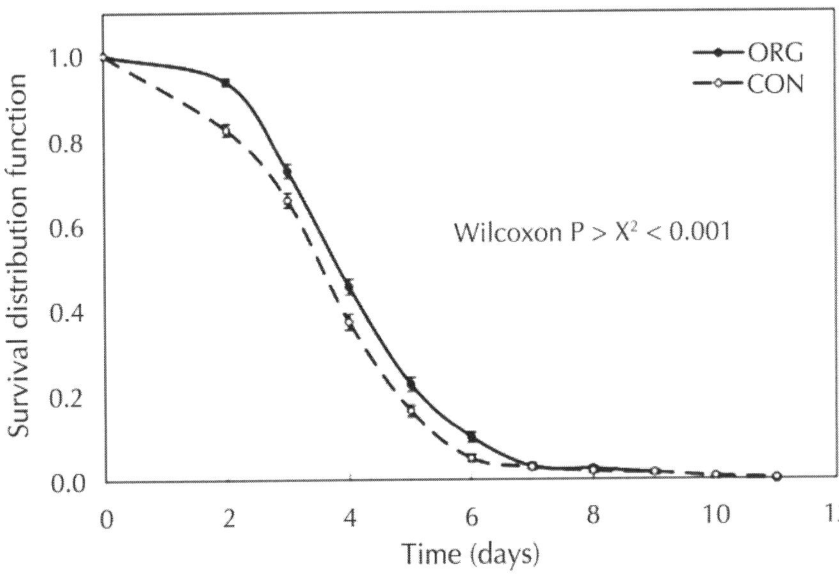

FIGURE 1 Survival distribution curves of rot tests for 'Diamante' and 'San Juan' strawberry fruit sampled from the 5 pairs of ORG and CON farm fields in June and September 2004. Mean survival days were CON = 4.15 ± 0.06 and ORG = 4.54 ± 0.06. (Error bars indicate standard error.).

TABLE 2 Fruit characteristics (mean ± standard error) of strawberries from ORG and CON farms (n = 13).

Fruit Quality Variable (units)	ORG	CON	*P* Value
Fruit fresh weight (g)	24.07 ± 0.68	27.78 ± 0.68	0.001
Dry matter (%)	10.03 ± 0.20	9.26 ± 0.20	0.006
Fruit weight loss (%)	25.40 ± 5.16	27.52 ± 5.16	0.048
Fruit firmness (N)	4.36 ± 1.90	4.17 ± 1.90	0.30
External L* (+60 to −60)	37.66 ± 0.76	38.65 ± 0.76	0.030
External C* (+60 to −60)	42.21 ± 0.37	41.76 ± 0.37	0.25
External hab (°)	31.26 ± 0.63	32.14 ± 0.63	0.048
Total antioxidant activity (mmol Trolox equivalents/g FW)	11.88 ± 0.35	10.95 ± 0.35	0.019
Total phenolics (mg gallic acid equivalents/g FW)	1.37 ± 0.13	1.24 ± 0.13	0.0003
Total ascorbic acid* (mg/g FW)	0.621 ± 0.015	0.566 ± 0.015	0.009
Total anthocyanins (μg P-3-Glc† equivalents/g FW)	205 ± 19.4	192 ± 19.4	0.103

Fruit characteristics (mean ± standard error) of strawberries from ORG and CON farms (n = 13).
Strawberries ("Diamante," "Lanai," and "San Juan") were sampled from 13 pairs of ORG and CON farm fields in June and September 2004 and April, June, and September 2005. *Based on Dietary Reference Intakes (DRI) [75], a standard serving (140 g) [76] of the fresh ORG strawberries would supply 9–10% more of the daily vitamin C (ascorbic acid) requirement of adult men and women than would the CON strawberries.

††Pelargonidyn-3-glucoside.

Strawberries from ORG farms were significantly smaller (by 13.4%) than those from CON farms but had significantly greater dry matter content (by 8.3%) (Table 2). Fruit firmness and external color intensity (C*) were similar between CON and ORG berries but ORG berries were darker red (significantly lower L* and hab) than CON berries. Although their darker red color did not result in a preference for the appearance of ORG over CON "Lanai" and "San Juan" strawberries by consumer-sensory panels, these panels did prefer the appearance of ORG 'Diamante' berries to their CON counterparts (Table 3).

TABLE 3 Consumer sensory evaluations (mean±standard error) of strawberries on a 9 point hedonic/intensity scale from ORG and CON farms (n = 13).

Sensory Property	'Diamante'		'Lanai'		'San Juan'		*P* Value
	ORG	CON	ORG	CON	ORG	CON	
	Hedonic/intensity ratings						
Overall acceptance	6.09 a ± 0.23	5.35 b ± 0.23	6.24 a ± 0.29	6.24 a ± 0.29	6.09 a ± 0.27	6.36 a ±0.27	0.029
Flavor	5.95 a ± 0.16	5.17 b ± 0.16	6.08 a ± 0.17	5.92 a ± 0.17	5.86 a ± 0.19	6.07 a±0.19	0.044
Sweetness	5.56a ± 0.22	4.73 b ± 0.22	5.69 a ± 0.24	5.56 a ± 0.24	5.52 a ± 0.25	5.74 a±0.25	0.029
Appearance	6.73 a ± 0.37	5.97 b ± 0.37	6.78 a ± 0.39	6.97 a ± 0.39	7.09 a ± 0.39	7.03 a±0.39	0.067
	ORG			**CON**			
Juiciness	6.21 ± 0.09			6.35 ± 0.09			0.11
Tartness	4.61 ± 0.27			4.75 ± 0.27			0.38

Consumer sensory evaluations (mean±standard error) of strawberries on a nine-point hedonic/intensity scale from ORG and CON farms (n = 13). Strawberry fruit ("Diamante," "Lanai," and "San Juan") were sampled from 13 pairs of organic and CON farm fields in September 2004 and April, June, and September 2005. Differences between values within rows followed by different letters are significant at P< 0.05.

TABLE 4 Concentration of specific polyphenols (mean ± standard error) in strawberry fruit from ORG and CON farms (n = 13).

Polyphenol (mg 100 g^{-1} FW)	'Diamante'		'Lanai'		'San Juan'		*P* Value
	ORG	CON	ORG	CON	ORG	CON	
	April						
Quercetin glycoside	4.00 ± 1.38	6.72* ± 1.38	9.18† ± 1.41	5.43 ± 1.41	9.01 ± 1.56	7.60±1.56	0.009
Quercetin, total	7.02 ± 1.17	9.45* ± 1.17	11.71† ± 1.17	7.92 ± 1.17	11.22 ± 1.56	10.11±1.56	0.020
Kaempferol	0.93 ± 0.08	1.13† ± 0.08	0.99 ± 0.08	1.05 ± 0.08	1.28* ± 0.10	1.07±0.10	0.026
	June						
Quercetin glycoside	6.27 ± 1.17	7.20 ± 1.17	2.87 ± 1.41	6.09* ± 1.41	5.01 ± 1.28	5.32±1.28	0.009
Quercetin, total	8.78 ± 1.14	9.72 ± 1.14	6.42 ± 1.17	8.80* ± 1.17	7.92 ± 1.15	7.81±1.15	0.020
Kaempferol	1.21‡ ± 0.07	0.96 ± 0.07	0.98 ± 0.08	1.03 ± 0.08	1.06 ± 0.07	0.98±0.07	0.026
	September						
Quercetin glycoside	4.97 ± 1.17	4.87 ± 1.17	3.89 ± 1.41	3.93 ± 1.41	4.90 ± 1.28	7.13*±1.28	0.009
Quercetin, total	7.51 ± 1.14	7.33 ± 1.14	6.61 ± 1.17	6.57 ± 1.17	7.37 ± 1.15	9.19±1.15	0.020
Kaempferol	0.96 ± 0.07	0.92 ± 0.07	1.03 ± 0.08	1.05 ± 0.08	0.93 ± 0.07	1.00±0.07	0.026

TABLE 4 *(Continued)*

	ORG	CON	
Quercetin	2.79 ± 0.06	2.71 ± 0.06	0.17
Kaempferol glycoside	4.28 ± 0.97	4.34 ± 0.97	0.88
Kaempferol, total	5.32 ± 1.02	5.35 ± 1.02	0.93
Ellagic acid glycoside	55.0 ± 13.1	53.8 ± 13.1	0.92
Ellagic acid	2.27 ± 1.48	2.08 ± 1.48	0.70
Ellagic acid, total	57.2 ± 1.31	55.9 ± 1.31	0.88
Phloridzin glycoside	2.04 ± 0.29	2.24 ± 0.29	0.49
Phloretin	2.40 ± 0.04	2.43 ± 0.04	0.56
Phloretin, total	4.42 ± 0.31	4.64 ± 0.31	0.41
R-Naringin glycoside	2.90 ± 0.95	1.35 ± 0.95	0.27
S-Naringin glycoside	2.90 ± 0.98	1.46 ± 0.98	0.32
R-Naringenin	0.43 ± 0.07	0.44 ± 0.07	0.83

TABLE 4 *(Continued)*

S-Naringenin	0.24 ± 0.07	0.29 ± 0.07	0.51
R-Naringenin, total	3.31 ± 0.95	1.77 ± 0.95	0.28
S-Naringenin, total	3.12 ± 0.98	1.73 ± 0.98	0.34

Concentration of specific polyphenols (mean ± standard error) in strawberry fruit from ORG and CON farms (n = 13).
Fruit were sampled in June and September 2004 and April, June, and September 2005 from 13 pairs of ORG and CON farm fields. Least square means ± standard error of the means. Probabilities (P values) for treatment x variety x month interactions for quercetin glycoside, total quercetin, and kaempferol, and for treatment main effects for the remaining polyphenols are given.
*Means are notably different at $P < 0.10$.
††Means are significantly different at $P < 0.05$.
‡‡Means are significantly different at $P < 0.01$.

The ORG strawberries had significantly higher total antioxidant activity (8.5% more), ascorbic acid (9.7% more), and total phenolics (10.5% more) than CON berries (Table 2), but significantly less phosphorus (13.6% less) and potassium (9.1% less) (Table 1). Specific polyphenols, such as quercetin and ellagic acid, showed mixed or no differences (Table 4). Strawberries are among the most concentrated sources of vitamin C and other antioxidant compounds in the human diet [32]. Dietary antioxidants, including ascorbic acid (i.e., vitamin C) and phenolic compounds offer significant potential human health benefits for protection against diseases [33, 34]. For example, Olsson et al. [35] reported decreased proliferation of breast and colon cancer cells by extracts of organically grown strawberries compared to CON berries, with ascorbic acid concentrations correlated negatively with cancer cell proliferation. Although the greater potassium concentration in the CON strawberries is a plus, strawberries are not among the richest sources of potassium or even phosphorus [36]. Interestingly, less phosphorus in the diet may be considered desirable, given the negative effects of the increasing U.S. consumption of phosphorus [37] on vitamin D and calcium metabolism [38], and the resulting potential risk to bone health.

Using hedonic/intensity ratings, consumer-sensory panels found ORG "Diamante" strawberries to be sweeter and have preferable flavor, appearance, and overall acceptance compared to CON "Diamante" berries (Table 3). The ORG and CON "Lanai" and "San Juan" berries were rated similarly. Sensory results of sweeter tasting "Diamante" strawberries were confirmed by higher soluble solids content measured in the laboratory (Table 5).

12.3.2 Soil Quality

Soils were sampled and analyzed from the top (0–10 cm) and bottom (20–30 cm) of the raised mounds in June 2004 and 2005. The organically managed surface soils compared to their CON counterparts contained significantly greater total carbon (21.6% more) and nitrogen (30.2% more) (Table 6). ORGmatter (total carbon) can have a beneficial impact on soil quality, enhancing soil structure and fertility and increasing water infiltration and storage [39]. Levels of extractable nutrients were similar in the two systems, with the exception of zinc, boron, and sodium being significantly higher and iron being notably higher in the organically farmed surface soils.

Organically managed surface soils also supported significantly greater microbial biomass (159.4% more), microbial carbon as a percent of total carbon (66.2% greater), readily mineralizable carbon (25.5% more), and microbial carbon to mineralizable carbon ratio (86.0% greater) (Table 6). These indicate larger pools of total, labile, and microbial biomass C and a higher proportion of soil total and labile C as microbial biomass. All measures of microbial activity were significantly greater in the organically farmed soils, including microbial respiration (33.3% more), dehydrogenase (112.3% more), acid phosphatase (98.9% more), and alkaline phosphatase (121.5% more). The organically farmed soils had a significantly lower qCO_2 metabolic quotient, indicating that the microbial biomass in the organically farmed soils was 94.7% more efficient or under less stress than in the conventionally farmed soils [40]. These same differences, except for qCO_2, alkaline phosphatase, iron, boron, and sodium, were also observed in soils from the bottom of the mounds (20–30 cm depth). To quantify soil microbial

TABLE 5 Soluble solids, titratable acidity (TA), soluble solids/TA ratio, reducing sugars, total sugars, and pH (mean ± standard error) of strawberry fruit from ORG and CON farms (n = 13).

Variable	'Diamante'		'Lanai'		'San Juan'		*P* Value
	ORG	**CON**	**ORG**	**CON**	**ORG**	**CON**	
Soluble solids (°brix)	8.97 a ± 0.48	7.68 b ± 0.48	8.98 a ± 0.58	9.52 a ± 0.58	8.96 a±0.53	8.71 a±0.53	0.091
TA (mg citric acid g^{-1} FW)	9.16 a ± 0.20	7.52 bc ± 0.20	7.18 bc ± 0.26	7.51 bc ± 0.26	6.97 c±0.24	7.79 b±0.24	0.0005
		ORG			CON		
Soluble solids/TA		1.16 ± 0.07			1.14 ± 0.07		0.62
Reducing sugars (mg Glc g^{-1} FW)		69.1 ± 2.05			69.4 ± 2.05		0.93
Total sugars (mg Glc g^{-1} FW)		73.0 ± 2.38			78.4 ± 2.38		0.13
pH		3.77 ± 0.06			3.81 ± 0.06		0.105

Soluble solids, titratable acidity (TA), soluble solids/TA ratio, reducing sugars, total sugars, and pH (mean ± standard error) of strawberry fruit from ORG and CONfarms (n = 13).

Fruit were sampled from 13 pairs of organic and CON farms in June and September 2004 and April, June, and September 2005. Probabilities (P values) for treatment x variety interactions for soluble solids and TA and for treatment main effects for the remaining variables are given. Differences among treatments within each row followed by different letters are significant at $P < 0.05$.

TABLE 6 Soil properties (mean ± standard error) at two depths (0–10 and 20–30 cm) from ORG and CON strawberry farms (n = 13)

Soil Property	ORG (0–10 cm)	CON (0–10 cm)	*P* Value	Organic (20–30 cm)	Conventional (20–30 cm)	*P* Value
Sand (g 100 g^{-1} soil)	60.3 ± 7.9	60.5 ± 8.2	0.931	61.0±8.0	59.8±8.7	0.644
Silt (g 100 g^{-1} soil)	26.8 ± 4.8	26.4 ± 5.5	0.619	25.8±4.9	27.3±5.7	0.821
Clay (g 100 g^{-1} soil)	13.0 ± 3.4	13.1 ± 2.9	0.925	13.2±3.5	12.9±3.2	0.384
Nitrate (mg kg^{-1} soil)	46.8 ± 12.1	31.6 ± 7.3	0.402	24.5±3.9	22.9±7.0	0.866
Ammonium (mg kg^{-1} soil)	2.8 ± 0.3	2.9 ± 0.3	0.105	2.5±0.2	2.7±0.3	0.316
Phosphorus (mg kg^{-1} soil)	60.9 ± 13.3	64.5 ± 7.7	0.652	60.1±13.5	72.1±10.4	0.173
Sulfur (mg kg^{-1} soil)	134 ± 30	119 ± 38	0.76	119±39.4	55.7±14.8	0.140
Boron (mg kg^{-1} soil)	0.88 ± 0.23	0.74 ± 0.25	0.043	0.71±0.19	0.75±0.30	0.441
Zinc (mg kg^{-1} soil)	2.88 ± 0.37	1.97 ± 0.12	0.048	2.42±0.37	1.81±0.24	0.097
Manganese (mg kg^{-1} soil)	4.52 ± 0.50	7.64 ± 2.01	0.217	3.13±0.37	3.68±0.47	0.196
Copper (mg kg^{-1} soil)	1.37 ± 0.31	1.17 ± 0.25	0.216	1.40±0.31	1.23±0.29	0.291
Iron (mg kg^{-1} soil)	28.6 ± 3.9	26.8 ± 5.0	0.064	26.4±3.4	31.4±5.4	0.203
Potassium (cmol kg^{-1} soil)	0.6 ± 0.1	0.5 ± 0.1	0.194	0.6±0.1	0.5±0.1	0.230
Calcium (cmol kg^{-1} soil)	10.7 ± 2.3	9.7 ± 2.1	0.165	10.3±2.5	9.6±2.2	0.519

TABLE 6 *(Continued)*

Soil Property	ORG (0–10 cm)	CON (0–10 cm)	*P* Value	Organic (20–30 cm)	Conventional (20–30 cm)	*P* Value
Magnesium (cmol kg^{-1} soil)	4.1 ± 1.1	4.2 ± 1.3	0.722	3.9 ± 1.2	4.20 ± 1.3	0.695
Sodium (cmol kg^{-1} soil)	0.4 ± 0.1	0.3 ± 0.1	0.001	0.3 ± 0.04	0.3 ± 0.04	0.858
Total bases (cmol (+) kg^{-1})	15.8 ± 3.5	14.7 ± 3.3	0.244	15.1 ± 3.7	14.6 ± 3.5	0.841
pH	7.05 ± 0.11	7.09 ± 0.16	0.953	7.16 ± 0.10	7.09 ± 0.17	0.694
Buffer capacity pH	7.51 ± 0.02	7.51 ± 0.03	0.908	7.53 ± 0.02	7.52 ± 0.03	0.789
EC (mmhos cm^{-1})	2.72 ± 0.34	2.18 ± 0.39	0.071	2.13 ± 0.37	1.50 ± 0.22	0.306
Total carbon (g kg^{-1} soil)	10.04 ± 0.15	8.25 ± 0.12	0.036	9.43 ± 0.17	7.71 ± 0.13	0.034
Total nitrogen (g kg^{-1} soil)	0.867 ± 0.014	0.666 ± 0.010	0.009	0.783 ± 0.015	0.625 ± 0.012	0.010
Readily mineralizable carbon (µg MinC g^{-1} soil)	17.7 ± 1.1	14.1 ± 1.2	0.009	14.9 ± 1.6	11.2 ± 1.2	0.019
Microbial biomass (µg MicC g^{-1} soil)*	249 ± 22.5	96 ± 6.8	0.000	211 ± 20.5	101 ± 12.1	0.042
MicC (% of total carbon)*	2.21 ± 0.13	1.33 ± 0.26	0.005	2.16 ± 0.31	1.54 ± 0.37	0.041
MicC $MinC^{-1}$*	16.0 ± 1.8	8.6 ± 0.6	0.004	16.8 ± 3.1	9.3 ± 0.5	0.049

TABLE 6 *(Continued)*

Soil Property	ORG (0–10 cm)	CON (0–10 cm)	*P* Value	Organic (20–30 cm)	Conventional (20–30 cm)	*P* Value
Basal respiration (μg CO_2-C g^{-1} soil ha^{-1})*	0.472 ± 0.055	0.354 ± 0.032	0.009	0.731 ± 0.186	0.348 ± 0.111	0.009
Dehydrogenase (μg TPF g^{-1} soil)	1.38 ± 0.21	0.65 ± 0.05	0.000	0.89 ± 0.14	0.52 ± 0.05	0.000
Acid phosphatase (μg p-nitrophenol g^{-1} soil)	121.5 ± 14.1	58.2 ± 5.6	0.009	104.7 ± 37.4†	53.1 ± 9.1†	0.039
Alkaline phosphatase (μg p-nitrophenol g^{-1} soil)	122.3 ± 13.0	55.6 ± 8.5	0.002	84.4 ± 17.0†	47.0 ± 13.0†	0.262
qCO_2 (μg CO_2-C ha^{-1} mg^{-1} MicC)*	1.9 ± 0.18	3.7 ± 0.33	0.003	3.5 ± 0.73	3.4 ± 0.67	0.838
Protease native (μg amino acid-N g^{-1} soil ha^{-1})	2.41 ± 0.29	2.81 ± 0.36	0.446	2.08 ± 0.29†	1.25 ± 0.35†	0.107
Protease potential (μg amino acid-N g^{-1} soil ha^{-1})	4.06 ± 0.65	3.49 ± 0.32	0.369	3.21 ± 0.25†	2.78 ± 0.31†	0.150
Mycorrhizae total colonized root length (mm)*	122 ± 11	104 ± 10	0.164	–	–	–

Soil properties (mean ± standard error) at two depths (0–10 and 20–30 cm) from ORG and CON strawberry farms (n = 13).
Soil samples were taken in June 2004 and June 2005, except where noted.
*Measured in June 2005 only.
††Measured in June 2004 only.

gene presence and diversity, we used a gene array termed GeoChip containing more than 24,000 oligonucleotide (50 mer) probes and covering 10,000 genes involved in nitrogen, carbon, sulfur, and phosphorus transformations and cycling, metal reduction and resistance, and ORG xenobiotic degradation [27], [41]. Microarray genes were analyzed both individually and within functional groups in soil samples from ORG and CON strawberry fields [42]. A functional group is a group of genes involved in a certain function or biogeochemical process in the soil. In this study, the following 11 functional groups were targeted: nitrogen fixation, nitrification, denitrification, sulfite reduction, pesticide degradation, other ORG xenobiotic degradation, metal reduction and resistance, and genes for the enzyme classe's dehydrogenase, urease, cellulase, and chitinase. Mean DNA microarray signal intensity of total detected genes was significantly greater in organically managed soils than in conventionally managed soils (Table 7) [42]. Similarly, mean SIs for the 11 gene functional groups were all significantly greater in organically managed soils (Table 7). The SIs of more than 32% (553) of 1711 individual genes detected were significantly higher in organically managed soils, while not, one was significantly higher in conventionally managed soils (Figure 2). The SIs is correlated with gene copy number and dependent on DNA labeling efficiency [26]. Mean labeling efficiency of DNA from ORG and CON soils was similar (1.23 and 1.25 μmol Cy 5/μl DNA solution, respectively, P = 0.78), demonstrating that the detected differences were not introduced by differing labeling efficiencies and that functional genes and likely the organisms that carry them were more abundant in organically managed soils.

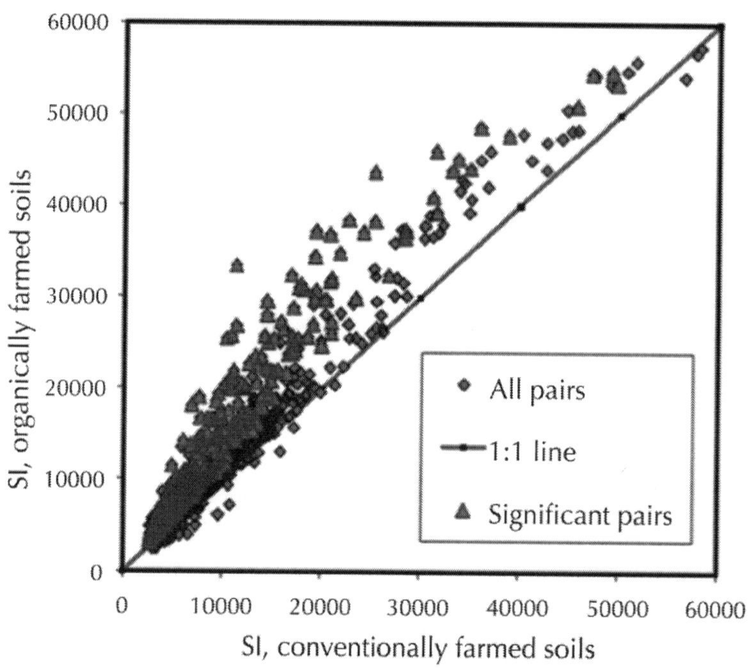

FIGURE 2 A scatter plot of (SIs) of 1711 individual genes on GeoChip microarrays.

TABLE 7 Soil DNA microarray (SIs) and diversity (mean ± standard error) of total detected genes and gene functional and organism groups from ORG and CON strawberry farms (n = 8).

Soil Functional Group or Organism Group	SIs (103)			Diversity (Simpson's Reciprocal Index)		
	ORG	CON	*P* Value	ORG	CON	*P* Value
Total detected genes	13479 ± 874	9350 ± 1003	0.008	656 ± 31	504 ± 34	0.015
N fixation	744 ± 59	547 ± 76	0.018	44 ± 1	38 ± 2	0.034
Nitrification	262 ± 12	201 ± 13	0.004	7 ± 0.3	6 ± 0.3	0.012
Denitrification	552 ± 46	405 ± 59	0.029	33 ± 2	26 ± 3	0.010
Sulfite reduction	529 ± 36	368 ± 40	0.009	37 ± 1	31 ± 1	0.004
Pesticide degradation	1970 ± 131	1322 ± 143	0.006	104 ± 4	81 ± 4	0.004
Other ORG xeno-biotic degradation	3999 ± 253	2819 ± 296	0.008	193 ± 10	146 ± 11	0.012
Metal reduction and resistance	2580 ± 164	1750 ± 164	0.008	112 ± 5	84 ± 5	0.010
Dehydrogenase	171 ± 9	118 ± 11	0.004	7 ± 0.3	6 ± 0.3	0.245
Urease	621 ± 48	394 ± 59	0.008	36 ± 3	28 ± 3	0.031
Cellulase	819 ± 60	569 ± 65	0.012	51 ± 2	41 ± 2	0.012
Chitinase	347 ± 25	240 ± 23	0.016	16 ± 1	12 ± 0.7	0.024
Fungi	164 ± 9	108 ± 10	0.003	17 ± 4	13 ± 3	0.025
Prokaryotes	12818 ± 837	9088 ± 964	0.008	624 ± 29	487 ± 33	0.011
Fungi/Prokaryotes Ratio	0.013 ± 0.000	0.012 ± 0.000	0.323	0.027 ± 0.007	0.027 ± 0.006	0.343

Soil DNA microarray SIs and diversity (mean ± standard error) of total detected genes and gene functional and organism groups from ORG and CON strawberry farms (n = 8).
Functional groups from Reeve et al. [42].

Organically managed soils exhibited significantly more endemic genes ($P < 0.0001$); more specifically, 233 genes were detected only in the organically managed soils and 2 genes were detected only in conventionally farmed soils. Genetic diversity was also significantly greater in organically managed soils across all detected genes (total) and for 10 of the 11 functional groups (Table 7). Greater diversity within a functional group may simply be redundant, particularly at high levels of diversity [43]. Conversely, greater diversity may help support the resulting ecosystem function or biogeochemical process in a broader range of environmental conditions [44] or in changing environments [45]. The findings of greater enzyme activities in organically managed soils indicate a greater functional capacity. Greater functional gene abundance in organically managed soils indicates a larger functional population. Greater functional gene diversity in organically managed soils suggests that ORG systems may also support more stable or resilient ecosystem functioning. Some of the 11 functional groups addressed on the GeoChip are purely prokaryotic functions (e.g., nitrogen fixation, nitrification, and denitrification), while others are characteristics of both prokaryotes and eukaryotes (mainly fungi). To ensure that neither group biased the SIs and diversity results for either the ORG or CON farming systems, we separated out fungal and prokaryotic-derived genes into their respective groups and calculated the ratio of fungi to prokaryotes for gene SIs and diversity in the two agroecosystems. Not only are the fungi numbers higher for both SIs and diversity in the ORG agroecosystems, but also are the prokaryote numbers, too (Table 7). However, ratios of fungi/prokaryotes for either SIs or diversity are similar for the two farming systems, which dispel the concern that the data presented selectively favored prokaryotes over eukaryotes in either the ORG or CON agroecosystems. The large differences in soil microbial properties and soil functional gene abundance and diversity between the organically and conventionally farmed soils are most likely due to a combination of factors: chemical fumigation with methyl bromide of the conventionally farmed soils, lack of synthetic pesticide use on the ORG fields, and double the application rates of compost to the ORG fields compared to the CON fields. A number of studies have documented changes in microbial diversity due to fumigants and pesticides [46-48], although the majority of changes were short-term, with microbial populations generally returning to normal after a few weeks or months. Many of these studies were conducted using simulated agricultural fumigation in a laboratory, and nearly all used a single fumigation event with no regard to past history of fumigation or pesticide use on the studied soil. Two- to three-year field studies with annual fumigation have shown methyl bromide to alter some microbial properties and enzymatic functions but the effects were inconsistent [49, 50]. In this study, in which soil samples were taken about 5 to 6 months after fumigation, was conducted on ORG and CON fields with longer histories (at least 5 years) of both ORG and CON (with fumigation) management, likely contributing to the detection of some persistent effects on the microbial population. The ORG fields also received 20.2–24.6 Mg compost per hectare, almost twice the rate of the CON fields at 11.2–13.4 Mg compost per hectare. Mäder et al. [23] found that soil amendment with animal manures in ORG farming systems increased microbial biomass and enzyme activity and altered the structure of the microbial community. Crop rotations likely played a minor role in the differences in soil properties in study because rotations were similar for the ORG and CON agroecosystems; that is, the ORG and CON farms used 2 year rotations, in which strawberries were followed by broccoli, lettuce, or a cover crop in the 2nd year. The ORG strawberries and their soils were of higher quality compared to their CON counterparts. Specifically, the ORG strawberries, while having lower concentrations of phosphorus and potassium, had higher antioxidant activity and concentrations of ascorbic acid and phenolic compounds, longer shelf life, greater dry matter, and, for "Diamante," better taste and appearance. The organically farmed soils had more carbon and nitrogen, greater microbial biomass and activity, and greater functional gene abundance and diversity. This study demonstrates that soil DNA analyzes using microarray technology can be used as an additional measurement of soil quality. Sustainability study also demonstrates the benefits of using an interdisciplinary methodology that comprehensively and quantitatively compares numerous indices of fruit and soil quality from multiple, commercial ORG and CON farms, multiple varieties and soils, and multiple sampling times.

12.4 CONCLUSION

The findings show that the ORG strawberry farms produced higher quality fruit and that their higher quality soils may have greater microbial functional capability and resilience to stress. These findings justify additional investigations aimed at detecting and quantifying such effects and their interactions.

KEYWORDS

- **Microarray technology**
- **Signal intensity**
- **Signal-to-noise ratio**
- **Simpson's reciprocal index**
- **Strawberry**

AUTHORS' CONTRIBUTIONS

Conceived and designed the experiments: John P. Reganold, Preston K. Andrews, Jennifer R. Reeve, and Lynne Carpenter-Boggs. Performed the experiments: John P. Reganold, Preston K. Andrews, Jennifer R. Reeve, Lynne Carpenter-Boggs, Christopher W. Schadt, and Neal M. Davies. Analyzed the data: John P. Reganold, Preston K. Andrews, Jennifer R. Reeve, Lynne Carpenter-Boggs, and J. Richard Alldredge. Contributed reagents/materials/analysis tools: John P. Reganold, Preston K. Andrews, Christopher W. Schadt, Carolyn F. Ross, Neal M. Davies, and Jizhong Zhou. Wrote the paper: John P. Reganold, Preston K. Andrews, Jennifer R. Reeve, and Lynne Carpenter-Boggs. Set up and managed the overall study: John P. Reganold.

ACKNOWLEDGMENT

We thank Tom Sjulin of Driscoll's Strawberry Associates and Larry Eddings of Pacific Gold Farms for assistance with farm selection and fruit processing. We thank Linda

Klein, Amit Dhingra, Patricia Ericsson, David Huggins, Jeff Smith, and two anonymous reviewers for comments on drafts of this manuscript. We thank Jan Dasgupta, Marc Evans, Gregory Peck, Sean Swezey, Carolina Torres, Canming Xiao, and Jaime Yáñez for technical assistance.

REFERENCES

1. Willer, H., Rohwedder, M., and Wynen, E. Current statistics. *The World of Organic Agriculture—Statistics and Emerging Trends 2009*. H. Willer, L. and Kilcher (Eds.). Bonn: IFOAM, pp. 25–58 (2009).
2. Greene, C., Dimitri, C., Lin, B. H., McBride, W., Oberholtzer, L., et al. *Emerging Issues in the U.S. Organic Industry*. USDA Economic Research Service, Washington, D.C. (2009).
3. Lockie, S., Halpin, D., and Pearson, D. Understanding the market for organic food. *Organic Agriculture: A Global Perspective.* P. Kristiansen, A. Taji, J. Reganold (Eds.). CSIRO Publishing, Collingwood, pp. 245–258 (2006).
4. Baker, B. P., Benbrook, C. M., Groth, E. III, and Benbrook, K. L. Pesticide residues in conventional, integrated pest management (IPM)-grown and organic foods: insights from three US data sets. *Food Addit Contam*, **19**, 427–446 (2002).
5. Curl, C. L., Fenske, R. A., and Elgethun, K. Organophosphorus pesticide exposure of urban and suburban preschool children with organic and conventional diets. *Environ Health Persp.*, **111**, 377–382 (2003).
6. Lu, C., Toepel, K., Irish, R., Fenske, R. A., Barr, D. B., and Bravo, R. Organic diets significantly lower children's dietary exposure to organophosphorus pesticides. *Environ Health Persp.*, **114**, 260–263 (2006).
7. Drewnowski, A. Concept of a nutritious food: toward a nutrient density score. *Am J Clin Nutr*, **82**, 721–32 (2005).
8. Scalbert, A., Johnson, I. T., and Saltmarsh, M. Polyphenols: antioxidants and beyond. *Am J Clin Nutr*, **81**(suppl), 215S–217S (2005).
9. Soil Association. *Organic Farming, Food Quality and Human Health: A Review of the Evidence*. Soil Association, Bristol (2000).
10. Brandt, K. and Molgaard, J. P. Organic agriculture: does it enhance or reduce the nutritional value of plant foods? *J Sci Food Agr*, **81**, 924–931 (2001).
11. Worthington, V. Nutritional quality of organic versus conventional fruits, vegetables, and grains. *J Altern Complem Med*, **7**, 161–173 (2001).
12. Williams, C. M. Nutritional quality of organic food: shades of gray or shades of green? *Proc Nutrition Soc*, **61**, 19–24 (2002).
13. Magkos, F., Arvaniti, F., and Zampelas, A. Organic food: Nutritious food or food for thought? A review of the evidence. *Int J Food Sci Nut*r, **54**, 357–371 (2003).
14. Rembialkowska, E. Quality of plant products from organic agriculture. *J Sci Food Agric*, **87**, 2757–2762 (2007).
15. Benbrook, C., Zhao, X., Yáñez, J., Davies, N., and Andrews, P. *New Evidence Confirms the Nutritional Superiority of Plant-Based Organic Foods*. The Organic Center, Boulder (2008). Retrieved from www.organiccenter.org. (accessed on October 29, 2009).
16. Lairon, D. Nutritional quality and safety of organic food. A review. *Agron Sustain Dev*, **30**, 33–41 (2010).
17. Bourn, D. and Prescott, J. A comparison of the nutritional value, sensory qualities, and food safety of organically and conventionally produced foods. *Crit Rev Food Sci*, **42**, 1–34 (2002).
18. Dangour, A. D., Dodhia, S. K., Hayter, A., Allen, E., Lock, K., et al. Nutrional quality of organic foods: a systematic review. *Am J Clin Nutr*, **90**, 680–685 (2009).
19. Doran, J. W. and Parkin T. B. Defining and assessing soil quality. J. W. Doran, D. C. Coleman, D. F. Bezdicek, and B. A. Stewart (Eds.). Defining Soil Quality for a Sustainable Environment, *Soil Sci Soc Stet Am.*, Madison, pp. 3–21 (1994).

20. Granatstein, D. and Bezdicek, D. F. The need for a soil quality index: local and regional perspectives. *Am J Alt Agric*, **7**, 12–16 (1992).
21. Reganold, J. P., Palmer, A. S., Lockhart, J. C., and Macgregor, A. N. Soil quality and financial performance of biodynamic and conventional farms in New Zealand. *Science*, **260**, 344–349 (1993).
22. Reganold, J. P., Glover, J. D., Andrews, P. K., and Hinman, H. R. Sustainability of three apple production systems. *Nature*, **410**, 926–930 (2001).
23. Mäder, P., Fleissbach, A., Dubois, D., Gunst, L., Fried, P., et al. Soil fertility and biodiversity in organic farming. *Science*, **296**, 1694–1697 (2002).
24. Trewavas, A. A critical assessment of organic farming-and-food assertions with particular respect to the UK and the potential environmental benefits of no-till agriculture. *Crop Prot*, **23**, 757–781 (2004).
25. Ibekwe, A. M. Effects of fumigants on non-target organisms in soils. *Adv Agron*, **83**, 1–35 (2004).
26. Rhee, S. K., Liu, X., Wu, L., Chong, S. C., Wan, X., et al. Detection of genes involved in biodegradation and biotransformation in microbial communities by using 50-mer oligonucleotide microarrays. *Appl Environ Microb*, **70**, 4303–4317 (2004).
27. He, Z., Gentry, T. J., Schadt, C. W., Wu, L., Liebich, J., et al. GeoChip: A comprehensive microarray for investigating biogeochemical, ecological, and environmental processes. *ISME J*, **1**, 67–77 (2007).
28. Van Eijk, T. Holism and FSR. M. Collinson (Ed.) A History of Farming Systems Research. CABI Publishing, Wallingford, pp. 323–334 (2000).
29. California Strawberry Commission. Retrieved from www.calstrawberry.com (accessed on November 27, 2009).
30. Food and Agriculture Organization. FAOSTAT. Rome, United Nations. Retrieved from http://faostat.fao.org/site/567/default.aspx#ancor (accessed on November 11, 2009).
31. Adaskaveg, J. D., Gubler, D., Michailides, T., and Holtz, B. Efficacy and Timing of Fungicides, Bactericides, and Biologicals for Deciduous Tree Fruit, Nut, Strawberry, and Vine Crops 2009. University of California Online Statewide IPM Program. Retrieved from cemadera.ucdavis.edu/files/63413.pdf. (accessed on October 24, 2009).
32. Chun, O. K., Kim, D. O., Smith, N., Schroeder, D., Han, J. T., et al. Daily consumption of phenolics and total antioxidant capacity from fruit and vegetables in the American diet. *J Sci Food Agric*, **85**, 1715–1724 (2005).
33. World Health Organization and Food and Agriculture Organization. *Vitamin and Mineral Requirements in Human Nutrition 2nd ed.* WHO/FAO Publication, Bangkok (2004).
34. Halliwell, B., Rafter, J., and Jenner, A. Health promotion by flavonoids, tocopherols, tocotrienols, and other phenols: Direct or indirect effects? Antioxidant or not? *Am J Clin Nutr*, **81**, 268S–276S (2005).
35. Olsson, M. E., Andersson, C. S., Oredsson, S., Berglund, R. H., and Gustavsson, K. E. Antioxidant levels and inhibition of cancer cell proliferation in vitro by extracts from organically and conventionally cultivated strawberries. *J Agric Food Chem*, **54**, 1248–1255 (2006).
36. USDA Agricultural Research Service. USDA National Nutrient Database for Standard Reference, Release 22. Nutrient Data Laboratory Online (2009). Retrieved from http://www.ars.usda.gov/ba/bhnrc/ndl (accessed on June 2, 2010).
37. Standing Committee on the Scientific Evaluation of Dietary Reference Intakes, Food and Nutrition Board, Institute of Medicine Dietary Reference Intakes for Calcium, Phosphorus, Magnesium, Vitamin D, and Fluoride. National Academies Press, Washington, D.C. (1997).
38. Bringhurst, F. R., Demay, M. B., and Kronenberg, H. M. *Hormones and disorders of mineral metabolism*. J. D.Wilson, D. W.Foster, H. M. Kronenberg, and P. R. Larsen (Eds.). Williams Textbook of Endocrinology 9th ed. W. B. Saunders Company, Philadelphia, pp. 1155–1210 (1998).
39. Weil, R. R. and Magdoff, F. Significance of soil organic matter to soil quality and health. F. Magdoff and R. R.Weil (Eds.). *Soil Organic Matter in Sustainable Agriculture*. CRC Press, Boca Raton, pp. 1–43 (2004).

40. Smith, J. L. Soil quality: The role of microorganisms. G. Bitton (Ed.). *Encyclopedia of Environmental Microbiology*. Wiley, New York, pp. 2944–2957 (2002).
41. Zhou, J., Kang, S., Schadt, C. W., and Garten, C. T. Jr. Spatial scaling of functional gene diversity across various microbial taxa. *Proc Natl Acad Sci USA*, **105**, 7768–7773 (2008).
42. Reeve, J. R., Schadt, C. W., Carpenter-Boggs, L., Kang, S., Zhou, J., et al. Effects of soil type and farm management on soil ecological functional genes and microbial activities. *ISME J*, **42** (2010).
43. Andren, O., Bengtsson, J., Clarholm, M. Biodiversity and species redundancy among litter decomposers.. *The Significance and Regulation of Soil Biodiversity*. H. P. Collins, G. P. Robertson, and M. J. Klug (Eds.). Kluwer, Dordrecht, pp. 141–151 (1995).
44. Zak, D. R., Blackwood, C. B., and Waldrop, M. P. A molecular dawn for biogeochemistry. *Trends Ecol Evol*, **21**, 288–295 (2006).
45. Loreau, M., Naeem, S., Inchausti, P., Bengtsson, J., Grime, J. P., et al. Biodiversity and ecosystem functioning: Current knowledge and future challenges. *Science*, **294**, 804–808 (2001).
46. Zelles, L., Palojärvi, A., Kandeler, E., Von Lützow, M., Winter, K., et al. Changes in soil microbial properties and phospholipid fatty acid fractions after chloroform fumigation. *Soil Biol Biochem*, **29**, 1325–1336 (1997).
47. Engelen, B., Meinken, K., Von Wintzingerode, F., Heuer, H., Malkomes, H. P., et al. Monitoring impact of a pesticide treatment on bacterial soil communities by metabolic and genetic fingerprinting in addition to conventional testing procedures. *Appl Environ Microbiol*, **64**, 2814–2821 (1998).
48. Ibekwe, A. M., Papiernik, S. K., Gan, J., Yates, S. R., Yang, C. H., et al. Impact of fumigants on soil microbial communities. *Appl Environ Microbiol*, **67**, 3245–3257 (2001).
49. Klose, S. and Ajwa, H. A. Enzyme activities in agricultural soils fumigated with methyl bromide alternatives. *Soil Biol Biochem*, **36**, 1625–1635 (2004).
50. Stromberger, M. E., Klose, S., Ajwa, H., Trout, T., and Fennimore, S. Microbial populations and enzyme activities in soils fumigated with methyl bromide alternatives. *Soil Sci Soc Am J*, **69**, 1987–1999 (2005).
51. Guerena, M. and Born, H. Strawberries: Organic Production. Fayetteville, *ATTRA* (2007). Retrieved from www.attra.ncat.org/attra-pub/PDF/strawberry.pdf (accessed on November 25, 2009).
52. Gavlak, R., Horneck, D., Miller, R. O., and Kotuby-Amacher, J. *Soil, Plant, and Water Reference Methods for the Western Region*, 2nd ed. (WCC-103 Publication). Colorado State University, Fort Collins (2003).
53. Truswell, A. S., Bateson, D. J., Madafiglio, K. C., Pennington, J .A. T., Rand, W. M., et al. INFOODS guidelines for describing foods: A systematic approach to describing foods to facilitate international exchange of food composition data. *J Food Comp Anal*, **4**, 18–38 (1991).
54. McCance, R. A. and Widdowson, E. M. *The Composition of Foods, 6th Summary Edn.*, Royal Society of Chemistry, Cambridge (2002).
55. McGuire, R. Reporting of objective color measurements. *HortScience*, **27**, 1254–1255 (1992).
56. Arnao, M. B., Cano, A., and Acosta, M. The hydrophilic and lipophilic contribution to total antioxidant activity. *Food Chem*, **73**, 239–244 (2001).
57. Cano, A., Hernández-Ruíz, J., García-Cánovas, F., Acosta, M., and Arnao, M. B. An end-point method for estimation of the total antioxidant activity in plant material. *Phytochem Analysis*, **9**, 196–202 (1998).
58. Foyer, C. H., Rowell, J., and Walker, D. Measurements of the ascorbate content of spinach leaf protoplasts and chloroplasts during illumination. *Planta*, **157**, 239–244 (1983).
59. Andrews, P. K., Fahy, D. A., Foyer, and C. H. Relationships between fruit exocarp antioxidants in the tomato (Lycopersicon esculentum) high pigment-1 mutant during development. *Physiol Plantarum*, **120**, 519–528 (2004).
60. Singleton, V. L., Orthofer, R., and Lameula-Raventos, R. M. Analysis of total phenols and other oxidation substrates and antioxidants by means of Folin-Ciocalteu reagant. *Method Enzymol*, **299**, 152–179 (1999).

61. Woodward, J. R. Physical and chemical changes in developing strawberry fruits. *J Sci Food Agric*, **23**, 465–473 (1972).
62. Torres, C. A., Davies, N. M., and Yañez, J. A., and Andrews, P. K. Disposition of selected flavonoids in fruit tissues of various tomato (Lycopersicon esculentum Mill.) genotypes. *J Agric Food Chem*, **53**, 9536–9543 (2005).
63. Yañez, J. A. and Davies, N. M. Stereospecific high-performance liquid chromatographic analysis of naringenin in urine. *J Pharm Biomed Anal*, **39**, 164–169 (2005).
64. Branch, S. K. Guidelines from the International Conference on Harmonisation (ICH). *J Pharm Biomed Anal*, **38**, 798–805 (2005).
65. Boulanger, B., Rozet, E., Moonen, F., Rudaz, S., and Hubert, P. A risk-based analysis of the AAPS conference report on quantitative bioanalytical methods validation and implementation. *J Chromatog B*, **877**, 2235–2243 (2009).
66. Southgate, D. A. T. *Determination of Food Carbohydrates*. Applied Science Publishers, London (1976).
67. International Organization for Standardization (ISO). *Sensory Analysis—Guidelines for the Use of Quantitative Response Scales, 2nd edition*. (Reference No. ISO 4121:2003). ISO Geneva (2003).
68. Kalbfleisch, J. D. and Prentice, R. L. The Statistical Analysis of Failure Time Data. Wiley, New York (1980).
69. Anderson, J. P. E. and Domsch, K. H. A physiological method for the quantitative measurement of microbial biomass in soil. *Soil Biol Biochem*, **10**, 215–221 (1978).
70. Tabatabai, M. A. Soil enzymes. Methods of Soil Analysis. Part 2, Microbiological and Biochemical Properties. R. W. Weaver, S. Angle, P. Bottomley, D. Bezdicek, S. Smith, et al. (Eds.). *Soil Sci Soc Am*, Madison, pp. 775–833 (1994).
71. Ladd, J. N. and Butler, J. H. A. Short-term assays of soil proteolytic enzyme activities using proteins and dipeptide derivatives as substrates. *Soil Biol Biochem*, **4**, 19–30 (1972).
72. Sylvia, D. M. Vesicular-arbuscular mycorrhizal fungi. Methods of Soil Analysis. Part 2, Microbiological and Biochemical Properties. R. W. Weaver, S. Angle, P. Bottomley, D. Bezdicek, S. Smith, et al. (Eds.). *Soil Sci Soc Am*, Madison, pp. 351–378 (1994).
73. Zhou, J., Bruns, M. A., Tiedje, J. M. DNA recovery from soils of diverse composition. *Appl Environ Microbiol*, **62**, 316–322 (1996).
74. Wu, L., Liu, X., Schadt, C. W., and Zhou, J. Microarray-based analysis of subnanogram quantities of microbial community DNAs by using whole-community genome amplification. *Appl Environ Microbiol*, **72**, 4931–4941 (2006).
75. J. J. Otten, J. P. Hellwig, L. D. Meyers (Eds.). Dietary Reference Intakes: The Essential Guide to Nutrient Requirements. National Academies Press, Washington, D.C (2006).
76. U.S. Food and Drug Administration. Reference amounts customarily consumed per eating occasion, Title 21 CFR Section 101.12(b), (1994).

Index

O

T

U

V

W

Z

An environmentally friendly book printed and bound in England by www.printondemand-worldwide.com

PEFC Certified

This product is from sustainably managed forests and controlled sources

www.pefc.org

PEFC/16-33-415

MIX

Paper from responsible sources

FSC www.fsc.org FSC® C004959

This book is made entirely of sustainable materials; FSC paper for the cover and PEFC paper for the text pages.

#0186 - 300414 - C0 - 234/156/13 [15] - CB